SPACEFLIGHT

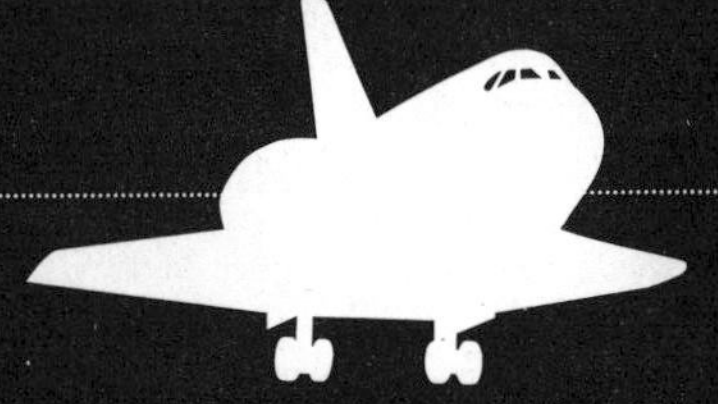

The MIT Press Essential Knowledge Series

Auctions, Timothy P. Hubbard and Harry J. Paarsch
The Book, Amaranth Borsuk
Carbon Capture, Howard J. Herzog
Cloud Computing, Nayan Ruparelia
Computing: A Concise History, Paul E. Ceruzzi
The Conscious Mind, Zoltan L. Torey
Crowdsourcing, Daren C. Brabham
Data Science, John D. Kelleher and Brendan Tierney
Extremism, J. M. Berger
Free Will, Mark Balaguer
The Future, Nick Montfort
GPS: A Concise History, Paul E. Ceruzzi
Haptics, Lynette A. Jones
Information and Society, Michael Buckland
Information and the Modern Corporation, James W. Cortada
Intellectual Property Strategy, John Palfrey
The Internet of Things, Samuel Greengard
Machine Learning: The New AI, Ethem Alpaydin
Machine Translation, Thierry Poibeau
Memes in Digital Culture, Limor Shifman
Metadata, Jeffrey Pomerantz
The Mind–Body Problem, Jonathan Westphal
MOOCs, Jonathan Haber
Neuroplasticity, Moheb Costandi
Open Access, Peter Suber
Paradox, Margaret Cuonzo
Post-Truth, Lee McIntyre
Robots, John Jordan
Self-Tracking, Gina Neff and Dawn Nafus
School Choice, David R. Garcia
Spaceflight: A Concise History, Michael J. Neufeld
Sustainability, Kent E. Portney
Synesthesia, Richard E. Cytowic
The Technological Singularity, Murray Shanahan
Understanding Beliefs, Nils J. Nilsson
Waves, Frederic Raichlen

SPACEFLIGHT

A CONCISE HISTORY

MICHAEL J. NEUFELD

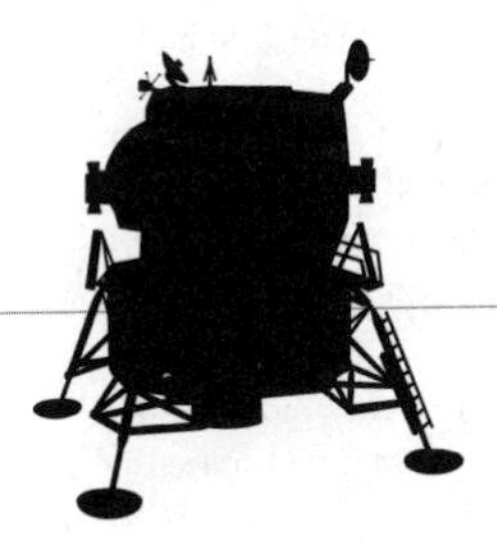

The MIT Press | Cambridge, Massachusetts | London, England

This book was set in Chaparral Pro by Toppan Best-set Premedia Limited. Printed and bound in the United States of America.

Library of Congress Cataloging-in-Publication Data

Names: Neufeld, Michael J., 1951- author.
Title: Spaceflight : a concise history / Michael J. Neufeld.
Other titles: Space flight
Description: Cambridge, MA : The MIT Press, [2018] | Series: The MIT Press essential knowledge series | Includes bibliographical references and index.
Identifiers: LCCN 2018013488 | ISBN 9780262536332 (pbk. : alk. paper)
Subjects: LCSH: Space flight--History. | Astronautics--United States--History. | Astronautics--Soviet Union--History. | Astronautics--Russia (Federation) | Space race--United States--History. | Space race--Soviet Union--History. | Manned space flight--History.
Classification: LCC TL788.5 .N48 2018 | DDC 629.4/109--dc23 LC record available at https://lccn.loc.gov/2018013488

10 9 8 7 6 5 4 3 2 1

CONTENTS

SERIES FOREWORD

The MIT Press Essential Knowledge series offers accessible, concise, beautifully produced pocket-size books on topics of current interest. Written by leading thinkers, the books in this series deliver expert overviews of subjects that range from the cultural and the historical to the scientific and the technical.

In today's era of instant information gratification, we have ready access to opinions, rationalizations, and superficial descriptions. Much harder to come by is the foundational knowledge that informs a principled understanding of the world. Essential Knowledge books fill that need. Synthesizing specialized subject matter for nonspecialists and engaging critical topics through fundamentals, each of these compact volumes offers readers a point of access to complex ideas.

Bruce Tidor
Professor of Biological Engineering and Computer Science
Massachusetts Institute of Technology

INTRODUCTION

Spaceflight is one of the greatest human achievements of the twentieth century. In 1900, only one or two persons in the world understood that the rocket could make space travel possible. Scarcely four decades later, German V-2 missiles began flying beyond the atmosphere. By 1963, the Soviet Union had launched the first satellites, hit the Moon, and put the first man and the first woman into Earth orbit. At the end of that decade, American astronauts orbited and then landed on the Moon. U.S. and Soviet robots reached the surface of Venus and Mars by the 1970s, and by 1989, American spacecraft had flown by all eight major planets. Four of those travelers were flung on one-way trips into interstellar space, the first human-made objects to leave not just Earth's gravitational influence, but even that of the Sun.

That direct exploration of the cosmos, in tandem with space-based and Earth-bound telescopes, has transformed human understanding of our planet, solar system, and universe. And yet exploration has been far from the only, or even dominant, reason we have gone into space. The great majority of spacecraft orbit Earth to provide services to, or gather information about, the planet. Since the 1960s, we have effectively annexed near-Earth space, from the twenty-four orbit at 22,200 miles (35,800 km) on down, and created a new zone of government and

Spaceflight is one of
the greatest human
achievements of the
twentieth century.

economic activity. What is done there is now essential to daily life, especially in the developed world, by providing global communications, satellite navigation, weather observation, military reconnaissance, missile early warning, Earth science, etc. The result is a growing but virtually invisible space infrastructure.

For many, "the space program" is still equated with human spaceflight. Yet astronauts have traveled less than about 400 miles (650 km) from Earth's surface since the last Apollo lunar mission in December 1972. That is expected to change in the 2020s, but whether that will lead to a vibrant future of Moon bases or Mars expeditions is an open question. And while important lessons have been learned in the nearly half century that spacefarers have only been in low Earth orbit (LEO), the rate of technological change, and of achievement, in human spaceflight pales in comparison to both deep-space robotic exploration and near-space infrastructure. Thus one of my primary purposes in this brief overview is to introduce educated lay readers to the full spectrum of activities that we humans have developed in space, and to the effects that has had on consciousness and culture.

A history also must describe origins and causes, not just the events and their effects. The human imagination and drive to explore certainly have something to do with spaceflight, but any activity out there costs a lot of money. Early in the twentieth century, amateur enthusiasm quickly gave way to international arms races and war as the primary drivers of development. Military and national security missions continue to be a very large part of what happens in Earth orbit. Garnering prestige and signaling technological strength has also been absolutely critical, especially for human spaceflight during the Cold War, but also afterward. And commercial competition and profit entered into the equation in the 1960s, at first only with communications satellites. Late in the Cold War, corporate activity began to expand into other sectors, and by the 2000s, even into human spaceflight.

Because nation-states have been the primary actors in this realm, space history is often written as a history of national programs, or of cooperative programs between states. That drives much of the narrative in this book, particularly in the first three chapters, which tell the story of the origins of space ideas and technology, the Cold War space race, and space science and exploration (which is a Cold War by-product). But transnational movements of people, ideas, and technology have always been part of the spaceflight story, and have grown in importance as corporations and new nations have taken on a larger role since the end of the Cold War. In the last three chapters, I discuss growth of a global space infrastructure (military and civilian), the rise of a global "astroculture," and the internationalization and privatization of human spaceflight after the Cold War. The global integration of world economies and political systems is the foundation for some of these changes, but spaceflight also impacts the process and rate of globalization, by influencing, among other things, the planet's culture and its networks of communication.

Barring a civilizational collapse induced by war or climate change, spaceflight appears to be here to stay, in large part because it has become so integral to life on Earth. Whether the human race will expand outward and become a multi-planet species, as spaceflight proselytizers have long predicted, is possible, but uncertain. Whatever happens, spaceflight by machines and humans is an astounding accomplishment with critical impacts on our lives. It behooves us to understand something of its history.

Acknowledgments

My NASM colleague David DeVorkin provided valuable comments on all of the chapters. Ron Doel, Matthew Shindell, Paul Ceruzzi, and Asif Siddiqi also gave helpful feedback. In addition, I would like to thank the anonymous reviewers of the MIT Press for their suggestions, and my editor, Katie Helke, for shepherding the contract, the review process, and the production of this book. My wife, Karen Levenback, gave love, support, and editorial suggestions throughout the process of writing, and our cats Pargiter and Ramsay were a welcome distraction during the hours spent at the computer.

Barring a civilizational collapse induced by war or climate change, spaceflight appears to be here to stay, in large part because it has become so integral to life on Earth.

1

SPACEFLIGHT DREAMS AND MILITARY IMPERATIVES

For millennia, the heavens were the realm of the gods and mythological beings. It was not a place one could imagine traveling to, with the possible exception of the Moon, whose visible face made it look like a real world. But even then, there was no non-supernatural way to go there. Creating a technology to escape Earth only gradually became imaginable in the eighteenth and nineteenth centuries, with the emergence of balloons, railroads, steamships, and other seemingly miraculous breakthroughs in transportation and communication. Another critical influence was the emergence of modern astronomy in Europe, which turned the Moon and the planets into places one could imagine walking on, even if going there seemed to be impossible.

Many fantastical and often satirical tales had been published featuring space travel, but none were as influential as those of French writer Jules Verne. His *From the Earth to the Moon* (1867) and its sequel *Around the Moon* (1870), like his other science fiction, set a new standard for technological realism. Notwithstanding the fact that the giant cannon he used to propel his spacefarers would have instantly flattened them at the moment of ignition, it inspired dreamers to imagine a flight to the Moon and how the problem could actually be solved.

One of them was Konstantin Tsiolkovsky, born in 1857, who overcame a hearing impairment to become a teacher in and around Kaluga in Tsarist Russia. In his spare time he was obsessed with inventing concepts for air and space flight. He spent more time on airship concepts than he did on space vehicles, but inspired by Verne, but also by a bizarre Russian philosophy called cosmism, in which the invention of space travel would lead to the perfection of mankind and reanimation of the dead, he also began searching for a way to propel something in a vacuum. By 1883 he realized that the rocket would work.[1]

Using rocketry for spaceflight was by no means obvious. The Chinese had invented the black-powder rocket around 1100 CE as a spin-off of gunpowder. It became important for fireworks and for warfare, particularly in Asia, and migrated to Europe in the late Middle Ages. Rockets got a second lease on life in the Napoleonic Wars, when the British inventor William Congreve developed iron casings and other refinements that made them competitive with conventional artillery. But rocketry faded again in the late nineteenth-century as rifled barrels allowed much greater accuracy. Again largely a firework, the rocket had no obvious potential to lift a vehicle to the almost unimaginable velocities required for spaceflight: 17,500 mph for low Earth orbit and 25,000 mph to escape the Earth, at a time when no human vehicle had yet traveled even 100 mph. Moreover, few understood that Isaac Newton's third law of motion, usually stated as "for every action there is an equal and opposite reaction" (as in the kickback of a gun), applied to the rocket. The popular fallacy was that its exhaust jet needed air to push against.

To see the rocket's potential, Tsiolkovsky made several imaginative leaps beyond the limitations of black powder. A self-taught amateur scientist, he used recent breakthroughs in chemistry to understand that burning two liquid fuels would greatly exceed its energy output and efficiency. Moreover, liquid oxygen (LOX) and liquid hydrogen would produce the almost highest possible energy output per unit mass of liquid propellants. (LOX, with a condensation point of –297°F, was first produced in a laboratory in 1883 and liquid hydrogen at –423°F in 1898.) Tsiolkovsky also used Newtonian mechanics to calculate the acceleration of such a hypothetical rocket, which lost mass as the propellant burned. He was the first to write down the basic equations of rocket motion. Coming out

Largely a firework, the rocket had no obvious potential to lift a vehicle to the almost unimaginable velocities required for spaceflight.

of these calculations, he eventually realized that the limitations of any one device could be exceeded by stacking them, now known as staging. That allowed deadweight to be discarded on the way up, making higher velocities feasible.

The eccentric teacher first published his ideas in turgid, didactic science fiction stories in the 1890s and as scientific papers in 1903 and 1911. But his publications were very obscure inside Russia until the eve of World War I, and entirely unknown outside it. In the early 1890s, an equally eccentric German inventor, Hermann Ganswindt, published his idea for a spaceship, which expelled individual masses to create a form of reaction propulsion, but his science was weak and his proposal was quickly forgotten. Around 1908, a new, younger group of theoreticians in Europe and America independently discovered many of Tsiolkovsky's ideas. Each of them was convinced that he was the first person on Earth to think of them.[2]

The spectacular flights of airplanes and airships in the early twentieth century spurred the imagination. If humans could fly in the atmosphere, what about beyond it? In one case, the Frenchman Robert Esnault-Pelterie, the connection was explicit—he was an important early aviator and aeronautical inventor who began to think about spaceflight as the next challenge. He published a paper laying out some of the theory in 1912, but he could only

Figure 1 Konstantin Tsiolkovsky, an eccentric Russian schoolteacher, was the first to work out the fundamental concepts and equations to prove that rocketry could make spaceflight possible. He began writing and publishing in the late nineteenth century, two decades ahead of other key theorists. *Source:* Smithsonian National Air and Space Museum (NASM 78-2461~P)

imagine that recently discovered atomic energy would have to be harnessed somehow to generate the velocities required. He missed the possibilities of the liquid-propellant rocket.

Of the early spaceflight theorists, two stand out as Tsiolkovsky's equals: Robert H. Goddard in the United States and Hermann Oberth in Central Europe. (Oberth was an ethnic German from Transylvania, part of Austria-Hungary until 1918 and Romania afterward.) As a seventeen-year-old student in Worcester, Massachusetts, Goddard became obsessed with spaceflight in 1899 after reading a newspaper serial resetting H. G. Wells's Martian invasion story, *War of the Worlds*, around Boston. Sitting in a cherry tree in his parents' backyard, Goddard had a vision of a craft ascending to Mars. Such was the quasi-religious nature of the experience that, when the great 1938 New England hurricane knocked over the tree, he wrote in his diary: "Cherry tree down. Have to carry on alone." His vision set off a private quest to find a propulsion system that could work in space; in early 1909, he realized it was the rocket and proceeded to work out all the principles. Shortly thereafter, Oberth, who was born in 1894, reached the same conclusions after being inspired by Verne. Son of a doctor and eventually a medical student, Oberth also made the earliest investigations of the possible effects of weightlessness. Less apparently eccentric than Goddard and Tsiolkovsky, he would end up later in life a full-blown crackpot, writing books about communicating telepathically with aliens. The topic of space travel seems to have attracted rather strange people. For most, it remained either remotely futuristic or completely lunatic.[3]

Goddard was one of the few theoreticians who became a successful rocket experimenter. Having survived a near-fatal bout of tuberculosis, he got a PhD in physics and became a junior professor at his alma mater, Clark University in Worcester. He began empirical experiments, including demonstrating that the rocket worked in a vacuum vessel. He experimentally derived how much flash powder would be needed to demonstrate that one of his rockets had impacted the night side of the Moon. In late 1916, he wrote to the Smithsonian Institution in Washington, DC, seeking more funding than he could get at Clark. He got lucky. Always very cautious about revealing his grand dream of spaceflight, he argued for his new rocket invention's ability to lift instruments into the upper atmosphere. It fit perfectly with the program of Charles Greeley Abbot, the Smithsonian Astrophysical Observatory's director, who wanted to measure the Sun's radiative output absent the atmosphere. In early 1917, with war raging in Europe, Goddard got a letter promising him $5,000 (worth twenty times that much today). When the United States entered World War I, he did tactical rocket work for the army. After the war, Abbot pressed him to finally publish a paper about his work and ideas.

The topic of space travel seems to have attracted rather strange people. For most, it remained either remotely futuristic or completely lunatic.

Reluctantly, the secretive Goddard finished *A Method of Reaching Extreme Altitudes*. Much of it was a dry mathematical treatise on basic principles, but at the end he discussed his idea for hitting the Moon with a rocket carrying flash powder. When the Smithsonian published the pamphlet in early January 1920, it also sent out a press release that included the lunar idea. It provoked an entirely unexpected reaction. American newspapers announced that a respectable college professor was planning to shoot a rocket to the Moon, a story that spread around the world. The press stoked rumors that Goddard was imminently going to launch himself. Volunteers wrote to join him on his lunar journey, while the *New York Times* ridiculed Goddard and the Smithsonian for lacking "the knowledge daily ladled out in high schools" that rockets needed air to push against. Despite the comic aspects of *A Method*'s release, it gave new public credibility to spaceflight and to rocketry as the way to get there.[4]

Rise of the Spaceflight Movement

In Germany and Austria, the chaos and revolution that followed the lost war obscured the Goddard story. But in 1923, Hermann Oberth published a small book in Munich: *The Rocket into Interplanetary Space*. He had dropped out of medical school, served as a medic in the war, and then wrote it as a doctoral dissertation in astronomy that the University of Heidelberg refused to accept. It included more advanced ideas about spaceflight technology than anything previously published outside Russia. Oberth, back home in Romania, got lucky too. His cause was taken up by Max Valier, an Austrian pilot and advocate of a popular pseudo-scientific theory that ice made up most of the universe. In 1924, Valier published magazine articles and a book popularizing Oberth's ideas. Nascent space enthusiasm spawned a small movement, with societies forming in Austria in 1926 and Germany in 1927.[5]

In the new Soviet Union, the disastrous war, followed by the Bolshevik Revolution and the Civil War, had thrown Konstantin Tsiolkovsky into desperate poverty. He was even briefly arrested by the Communists. His prewar notoriety in popular science works was forgotten. But in 1924, a combination of wild rumors about Goddard and the publication of Oberth's book stirred a revival of his reputation. Russian space enthusiasts seized on him as their homegrown hero. They reissued his prewar publications and he wrote more. Others working on spaceflight ideas since before the war came out of the woodwork, notably Fridrikh Tsander. While the world's first spaceflight society, formed in 1924, quickly fell apart, the utopian climate of early Communist Russia fostered space ideas even more strongly than a Central Europe in postwar upheaval.[6]

Back across the Atlantic, Goddard did public speaking and wrote articles to counter the press, but mostly he worked quietly on rockets at Clark University, funded by the Smithsonian. His prewar idea for an improved black-powder rocket using cartridges, essentially a machine gun firing blanks, never worked. In 1921 he switched to liquid propellants, which he had carefully mentioned only in the endnotes to *A Method*. He chose easily obtainable gasoline and liquid oxygen. After slow progress that began to frustrate Abbot, now head of the Smithsonian, he launched the world's first liquid-propellant rocket on a frigid March 16, 1926, in Auburn, Massachusetts. The flimsy-looking contraption flew 184 feet. But he told no one outside his immediate circle and Abbot. Goddard was always tinkering, filing patent applications, and hoping to perfect his rocket before he would reveal it to the world. It was not until a summer 1929 launch provoked a new round of press coverage that he got lucky again. Famed aviator Charles Lindbergh intervened with the Guggenheim Foundation, leading to a new, more elaborate round of rocket development in New Mexico in the 1930s.

Knowing almost nothing about Goddard's work, including his liquid-propellant experiments, European rocket groups formed around 1930, naively hoping to develop technology that could make spaceflight possible in a very short time. In Germany, a rocket and space fad broke out in 1928 after Max Valier briefly aligned himself with automobile heir Fritz von Opel. Together and separately, they carried out a series of spectacular, if technologically pointless, black-powder rocket stunts with cars, rail cars, ice sleds and gliders. Movie director Fritz Lang, famous for *Metropolis*, released a spaceflight movie, *Woman in the Moon*, in fall 1929, with Oberth and a young science writer, Willy Ley, as scientific advisors. Lang lured Oberth from Romania to Berlin, and then funded him to launch a liquid-fuel rocket for the movie premiere.[7]

Oberth hired Rudolf Nebel, a slippery ex-fighter pilot and engineer, to assist him, but Oberth was a hopeless inventor. After a nervous breakdown, he fled back to Romania at the end of 1929. Nebel secretly secured 5,000 marks from the German army, which had begun looking at the technology, to finish and launch Oberth's rocket. Nothing much came of it except some small-scale engine tests in July 1930 that Oberth participated in. But that fall Nebel created Rocketport Berlin in an abandoned munitions dump in the northern part of the city, midwifed by the army. His impoverished experimenter group became the main activity of the Society for Space Travel, which was in decline due to financial bungling and the onset of the Great Depression. In 1931 they launched their first, primitive liquid-fuel rockets. One episodic participant was an aristocratic engineering student, Wernher von Braun, born in 1912, who had become a spaceflight fanatic after attempting to read Oberth's mathematically challenging book as a thirteen-year-old student.[8]

In the Soviet Union, an amateur rocket group formed in Moscow in 1931, and a small solid-propellant institution in Leningrad (St. Petersburg) was given an expanded research agenda the same year. Tsander led the Moscow group, but he soon died. Sergei Pavlovich Korolev, an aeronautical engineer six years older than von Braun, became the group's effective leader. In parallel to the German, he would become an indispensable organizer in the military-dominated rocket world to come. In 1933, the Moscow group launched its first rockets, just as it was merged with Leningrad to form the world's first government research establishment for the development of liquid- and solid-propellant rocketry. It had a close connection to the Red Army. That year also saw the Nazi seizure of power, which, within a year, led to German army control over rocket development and the end of amateur experimentation and publicity.

The Military Takes Over

The consolidation of totalitarian power by Stalin and Hitler, and the resultant withdrawal of German and Russian spaceflight enthusiasts from the international scene, ushered in three decades in which military missile development dominated rocketry. It also disrupted the transnational network that had formed in the late 1920s to spread the gospel of spaceflight through liquid-propellant rockets. Two multilingual writers were central to this network, Willy Ley in Berlin and Nikolai Rynin in Leningrad. Through international correspondence and the circulation of publications, Ley and Rynin linked true believers and experimenters in Europe and in the United States, where the American Interplanetary Society formed in New York City in 1930. As the Germans and Russians became uncommunicative, what was left of the network shifted to the axis between New York and England, where the British Interplanetary Society (BIS) formed in 1933. Ley himself fled Nazi Germany in early 1935, assisted by contacts in the BIS and its American counterpart. For the first few years, he led a marginal existence as a freelance science writer and rocket experimenter in the New York area. Rynin would die in the Nazi siege of Leningrad.

The American Interplanetary Society began by carrying out liquid-fuel experiments in imitation of the Germans. Its first leader, science fiction writer and public relations man G. Edward Pendray, visited Berlin in 1931 and corresponded with Ley until the latter landed in the United States under Pendray's sponsorship. In the mid-to-late thirties, the group changed its name to the less eccentric sounding American Rocket Society (ARS) and stopped trying to fly rockets. Its emergent leadership of young engineers and technicians concentrated on small-scale engine development, which is what they could afford out of their own, rather barren pockets.[9]

Robert Goddard kept his distance from the New York group, literally and figuratively. Funded by the Guggenheim Foundation, he moved to Roswell, New Mexico, in mid-1930 and stayed there until 1942, except for a two-year interruption in 1932–1934 due to the Great Depression's impact on Guggenheim investments. Contrary to the neglected genius mythology that later grew up around Goddard, he was one of the best-funded scientists in the United States in the 1930s, at a time when private philanthropy still dominated a much smaller science and technology research sector. He made significant advances by 1935, with vehicles that reached near 10,000 feet and velocities of several hundred miles an hour. But his patrons again became frustrated when he never came close to delivering on multiple promises to reach high altitudes with an instrument-laden "sounding rocket," as it would be called after World War II. Abbot, Lindbergh, and Harry Guggenheim urged him to seek outside assistance in the late 1930s, notably from a new rocket group led by Frank Malina that began in 1936 at the California Institute of Technology (Caltech) in Pasadena. Goddard resisted; he was incapable of making the transition to leading massive rocket engineering teams, such as those that would emerge under von Braun and Korolev. He remained a tinkerer, determined to work with a handful of men sworn to secrecy. He could claim many liquid-fuel rocket firsts, yet his influence on later technological development would be almost zero. His greatest impact would always be inspiring others to believe in rockets as the way to spaceflight.[10]

As Goddard was reaching a technological dead end, a German army team was making fundamental breakthroughs. In late 1932, two months before Hitler became chancellor, the army hired twenty-year-old Wernher von Braun to write a secret doctoral dissertation on liquid-propellant rocketry. It was a very small beginning, but soon the project would experience a massive scale-up, thanks to Nazi rearmament money, von Braun's talent, and artillery officers' fascination with the "long-range rocket"—what we would call a ballistic missile—as a possibly decisive surprise weapon. In 1935 the army allied with the rapidly rising Luftwaffe (air force); in 1936 the two services began constructing a super-secret, joint rocket center at Peenemünde, on a Baltic island north of Berlin. Under the charismatic leadership of von Braun and his military superior, Lt. Col. (later Gen.) Walter Dornberger, the army rocket project soon hired gyroscope specialists, aerodynamicists, and chemical and mechanical engineers to make key breakthroughs in guidance and control, supersonic aerodynamics, and rocket engines.[11]

Their first military objective was a ballistic missile called the A-4, the fourth design in their rocket series, which was to carry a 1-metric-ton (2,200 lb.) high-explosive or poison-gas warhead at least 175 miles. Nazi propagandists later dubbed it Vengeance Weapon 2 or V-2. By 1939, Dornberger and von Braun's team had begun testing LOX/alcohol engines with the 25-metric-ton (55,000 lb.) thrust needed to loft it. Goddard and the Soviets had never gotten beyond a few hundred pounds of thrust. The German duo secured the funds to build the world's largest and fastest supersonic wind tunnel in Peenemünde, so as to master the aerodynamics of a vehicle that would approach five times the speed of sound. Creating inertial and radio guidance-and-control systems proved to be their most challenging task, resulting in a massive expansion of in-house and university expertise, plus more contracting with industrial firms. The army leadership, convinced they were going to get a superweapon, was willing to throw huge sums of money at the project, despite skepticism from Hitler, who cut back its construction priorities in 1940–1941. There is little evidence that his action slowed technology development, despite Dornberger's postwar scapegoating of the Führer for making the A-4/V-2 "too late" to change the course of the war.[12]

On October 3, 1942, the Peenemünde group succeeded on their third launch attempt; the missile went about 56 miles high and 120 miles downrange into the Baltic. It was the first man-made object to come near the edge of space, now conventionally defined as 100 km (62.1 mi.). It shattered all world records for a man-made object—range, velocity, and altitude. Von Braun and Dornberger prized this accomplishment as the first step into space, but they also used it to press the case for the missile's mass production as a weapon. Both were Nazis in different ways: von Braun as an opportunistic party member and reluctant SS officer, and Dornberger as an outspoken proponent of National Socialism, even if officers could not become party members.[13]

In late 1942, Armaments Minister Albert Speer got Hitler to authorize V-2 production, after Allied victories in North Africa and Russia signaled a decisive turn in the war. Producing such an exotic weapon, however, was difficult due to the desperate shortage of skilled workers, mainly because of the manpower-consuming ground war with the Soviet Union. The whole economy was dependent on forced and slave labor from the occupied territories. In spring 1943, the army rocket project and Speer's ministry decided to use SS concentration camp prisoners for unskilled and semiskilled positions, while routinizing production as much as possible. After the British Royal Air Force attacked Peenemünde in August 1943, the Nazi leadership decided that V-2 assembly would be concentrated in a mine near Nordhausen in central Germany. The so-called Mittelwerk (Central Works) would draw labor from a newly created Dora subcamp of Buchenwald. Over five thousand missiles were produced, but tens of thousands of prisoners suffered or died as a result, something Dornberger and von Braun knew firsthand. They bear some responsibility for the massive war crimes of V-2 production, even if the SS was the primary perpetrator.[14]

When the Allies became aware of the German army program in 1943–1944, they accelerated their investigations of rocketry. All powers had already developed solid-propellant bombardment rockets based on new double-base (nitrocellulose-nitroglycerine) powders that were more powerful and did not have black powder's telltale smoke trail. But now the United States and the Soviet Union began to pay attention to the potential of liquid-propellant ballistic missiles and, thanks to the Luftwaffe's development of the winged V-1 "buzz bomb," air-breathing cruise missiles as well.

In 1944, the U.S. Army Ordnance Department funded a new missile project at General Electric called Hermes and converted the Caltech rocket project, which had been developing "jet-assisted take-off" (JATO) rockets for the U.S. Army Air Forces, into the Jet Propulsion Laboratory (JPL). (The word "rocket" lacked respectability in America because of the explosion of space-themed science fiction stories, comics, and movies in the 1930s; see chapter 5.) World War II JATO development had also diverted Robert Goddard to work for the navy in Maryland, and spurred the development of the first two liquid-propellant rocket companies in the United States: Reaction Motors, a spin-off of the ARS in New York, and Aerojet, a spin-off of the Caltech group in Pasadena. All of this indigenous development created valuable native-born expertise for the Cold War to come, even though during the war, actual results were modest.[15]

The Soviet Union became aware of German missile work by spying on the Germans and the Allies, and by the Red Army occupation of V-2 testing sites in Poland. In August 1944, Stalin ordered the release of two key rocket experts in the Gulag concentration camp system, Sergei Korolev and Valentin Glushko, the latter a specialist in rocket engines. Both had been imprisoned in 1938, and almost died, because of the horrifying Great Purge that cost the lives of several million people. The rocket research institute's two most senior leaders had been shot. Stalin's paranoid purge would later be a convenient explanation for the Soviet program's failure to yield the breakthroughs of the German, but the rocket institute had been hobbled by infighting in the mid-1930s over solid vs. liquid propellants, winged vs. ballistic missiles, and other technological choices. One of the German liquid-fuel project's virtues was its laser-like focus on the long-range ballistic missile powered by LOX and alcohol, a propellant combination von Braun adopted from Oberth and Rocketport Berlin.[16]

Yet the irony was that, when the V-2 was finally deployed against Western European cities in September 1944, it was no "wonder weapon." Josef Goebbels's

Propaganda Ministry had applied that label to the V-1 and the V-2 (as they were now called), and to other technological miracles that were supposed to salvage a disastrous war. The air force's pulsejet-powered V-1, first launched against London in June 1944, at least was relatively cheap and diverted Allied air defenses into shooting it down. The army's V-2 cost ten times as much, was too complex to manufacture and launch in large numbers, and diverted fewer Allied resources, as it came in supersonically, making it impossible to intercept. Its high-explosive warhead—a poison-gas alternative was never finished—made an impressive hole in the ground at that velocity, but accuracy and reliability were poor, as was true of the V-1. Both could barely hit a giant urban area. Especially in the V-2's case, it was an extraordinarily expensive way to drop a ton of high explosives. By 1943–1944, Britain and America had perfected four-engine bomber technology and operations to the point that they could burn down whole cities and kill tens of thousands of people in one night—and that was before the atomic bomb suddenly appeared in August 1945. The V-weapons offensive was, by comparison, spectacular but strategically ineffective. German missile and rocket technology was too early to be militarily useful rather than too late to alter the war's outcome.[17]

But it had potential and presented a major target for Allied forces as they invaded the Reich in spring 1945.

As the war ended, a scramble for German manpower and technology began that presaged the Cold War. The United States was the biggest winner, but the USSR, Britain, and a resurrected France all got spoils. Spaceflight had nothing to do with why German technology was desirable in 1945, but it would enable access to space within a very short time.

The Cold War Missile Race and the First Steps into Space

Dornberger and von Braun surrendered to the U.S. Army in the Alps on May 2, three weeks after American units overran the Mittelwerk. Since much of the Peenemünde leadership was near one of those two locations, the United States could skim off the top people in the program. The Ordnance Department also decided to take a hundred V-2s back home for testing, although mostly parts were shipped for lack of nearly complete missiles. That operation was done hurriedly, as the underground complex was in the future Soviet zone of occupation.

When the Red Army conquered Peenemünde in early May, it found the place stripped by the German evacuation, but the Mittelwerk's production machinery and leftover missile components provided a foundation for mastering German rocket technology. Korolev and Glushko, now in officers' uniforms, were included in the inspection teams.

Spaceflight had nothing to do with why German technology was desirable in 1945, but it would enable access to space within a very short time.

They helped set up special rocket institutes in their occupation zone, to which willing Germans engineers and scientists were lured with better pay and amenities than the Americans were offering. Most of those Germans, including a few key Peenemünders, would be shipped off to Russia at gunpoint in October 1946.[18]

Over summer 1945, the U.S. government had already created a program to import German and Austrian expertise, best known by its later designation, Project Paperclip. Gen. Dornberger was handed over to the British as a prisoner of war, but Wernher von Braun was picked to head a rocket group at Fort Bliss, outside El Paso, Texas. About 125 arrived by early 1946 to help military personnel and Project Hermes engineers reassemble and launch V-2s from the nearby White Sands Proving Ground in New Mexico.[19]

Some of the Germans had first helped the British Army prepare and launch three V-2s from Germany's North Sea coast as a learning exercise. But the British government quickly decided that it could not afford a major rocket program on top of jet aircraft development; it imported only a couple of dozen Germans and Austrians from rocket programs. Meanwhile, the French slowly began to attract engineers, scientists, and technicians to its own projects. They constituted a small German rocket group at Vernon, France, that would become a foundation for the French missile and space program.[20]

News of the V-2 and the subsequent capture of its key personnel excited the old space enthusiasts and animated new interest in spaceflight in the public and in the military, especially in the United States. The true believers immediately saw that this rocket was the technological breakthrough they had been waiting for, whatever its failings as a weapon. Excitement intensified when the army began launching V-2s from White Sands, New Mexico, in mid-1946, carrying the first scientific instruments to near space. (The Germans had launched V-2s vertically over 100 miles in 1944, but they carried no instrumentation.) New enthusiasts in the navy and the air force (which became independent from the army in 1947) even started secret satellite and Moon projects. Yet they were soon canceled, and missile research cut back, as the United States demobilized, slashed its budget, and tried to revert to its historic pattern of a small peacetime military. It would take the Cold War's rapid emergence in the late 1940s to change that.

The United States and its allies soon felt directly threatened by the brutal Stalinization of Eastern Europe, Communist intervention in Western and southern Europe, and the victory of the Chinese Communists in 1949. But from the Soviet point of view, their massive advantage in ground forces had been checkmated by the U.S. atomic bomb and the ring of American bases around Soviet-dominated territory. To catch up technologically,

Figure 2 American soldiers and technicians prepare a German V-2 ballistic missile for launching in White Sands on May 10, 1946—the first successful firing in the United States. The V-2 was not a very effective weapon, but it was a revolutionary breakthrough in liquid-propellant rocket technology that greatly accelerated the appearance of the ICBM and the space booster. *Source:* National Archives and Records Administration (USAF 32865AC)

and create the possibility of striking back, Stalin ordered that the V-2, the U.S. B-29 bomber, and the Fat Man plutonium bomb (the type used against Nagasaki) be copied. Soviet rocket experts, headed by Sergei Korolev, were frustrated by the V-2 order, as they would have preferred to start afresh, but they had no choice but to obey. With the forced evacuation of the missile institutes in East Germany in October 1946, Soviet and German rocket experts accelerated their work on preparing captured V-2s for firing. Launches began on the steppe east of Stalingrad in October 1947. The Germans, led by Helmut Gröttrup from the Peenemünde guidance group, helped solve some serious problems, but were soon isolated from their Soviet counterparts. Most were cordoned off in a camp on an island in a northern Russian lake and set to work on future concepts, with steadily diminishing influence on Soviet ballistic missile work led by Korolev. Their isolation was in preparation for sending them back home, beginning in the early 1950s. In a paranoid, totalitarian state without an immigration tradition, it proved impossible to absorb the German rocketeers, in direct contrast to the United States.[21]

Von Braun and his group, who constituted about one-fifth of the Paperclip Germans imported early in the program, had been put to work developing a cruise missile for the army. The test version would be launched on a V-2. But the defense cutbacks of 1946–1947 left von Braun frustrated by the glacial progress toward a large missile that might also advance his spaceflight dream. He wrote a science fiction novel about a Mars expedition, with an elaborate mathematical appendix proving its feasibility, in the hope of convincing the public. Fiction was not his strongpoint, and only the appendix was later published. But as the Cold War heated up, the federal government gradually increased money for rocket development and made Project Paperclip a pathway to citizenship, sweeping the Nazi past of von Braun and others under the rug. In 1950, the army concentrated rocket development at Redstone Arsenal in Huntsville, Alabama, transferring the Germans and several thousand Americans there. In the middle of the move, Communist North Korea invaded the south, heightening anti-Communist hysteria and accelerating federal defense spending. Von Braun's group was redirected to a nuclear-armed, super-V-2 called Redstone. It played a critical role in the early space race.

However, one must not privilege only the V-2/Huntsville line of development, as von Braun's followers later did. Although German technology gave American rocket and missile development a crucial jumpstart, the Jet Propulsion Laboratory, Aerojet, Reaction Motors, and General Electric's rocket division grew because of World War II investments. The Naval Research Laboratory in Washington, DC, built scientific instruments for White Sands V-2 launches and decided to develop its own large sounding rocket, Viking, contracting the Martin aircraft firm in Maryland to assemble it and Reaction Motors to provide its engine. The new U.S. Air Force (USAF) invested in rocket engines, leading to the further expansion of Aerojet and Reaction Motors, and the creation of Rocketdyne Division of North American Aviation, America's premiere liquid-propellant rocket motor enterprise in the 1960s. As the air force became America's primary long-range nuclear force, its rocket funding soon eclipsed the older services. It was closely allied with large West Coast aircraft firms like Convair, Douglas, and Boeing, all of whom claimed a stake in the missile business. Large solid-propellant rockets with new, more energetic chemical compositions, which originated in experiments at JPL and Aerojet during the war, soon came to dominate missile propulsion because of ease of storage and launch. That powered the rise of Thiokol, Hercules Powder, and other chemical engineering firms as rocket manufacturers.[22]

A critical moment for the arms and space races came when both the United States and the Soviet Union decided to proceed with intercontinental ballistic missiles (ICBMs). The American military had focused on cruise missiles after 1945, as guiding a winged vehicle in the atmosphere looked easier than a ballistic missile, given the V-2's poor accuracy. It took several years before propulsion and control problems with air-breathing missiles disillusioned the services. As inertial guidance systems based on gyroscopic stabilized platforms improved, and nuclear warheads grew lighter and more powerful, it became simpler to hurl a bomb on a trajectory than to fly it autonomously for hours to the other side of the world. After the United States carried out the first thermonuclear "hydrogen bomb" test in fall 1952, weapons designers promised a rapid breakthrough to compact devices. The monstrous yield of these bombs, a thousand times more powerful than those used on Japan, made accuracy less important; they could devastate even when impacting miles off-target. President Dwight Eisenhower's administration, which took office in January 1953, approved the air force's Atlas ICBM project and made it top national priority in 1954–1955. In 1955, Eisenhower, despite his serious reservations about the growth of the federal budget, also agreed to two intermediate-range ballistic missiles (each with a 1650-mi. range) as a stopgap capability against the Soviets. The air force ICBM group under Gen. Bernhard Schriever created the Thor from elements of Atlas, while von Braun's group was tasked with developing the Jupiter missile for the army and the navy. All of these rockets, burning LOX and kerosene, became satellite launchers a couple of years later.[23]

These crash programs, especially Thor and Jupiter, were responses to a growing threat from Soviet ballistic missile development. Just like von Braun, Korolev exhibited a genius for inspiring and coordinating diverse design bureaus, production firms and military offices to focus on a clear goal: long-range, rocket-powered ballistic missiles. Funded primarily by Soviet army artillery, he copied the V-2 as the R-1, and then doubled its range with the longer R-2. Mikhail Yangel's design bureau developed the R-5 medium-range missile, first tested in 1953, and converted it into the nuclear-armed R-5M, a source of American alarm about its bases and allies in Western Europe and Asia. In 1955, in parallel to Atlas, a post-Stalin Soviet leadership dominated by Nikita Khrushchev approved the first missile that could directly attack America, Korolev's R-7 ICBM. Because his nuclear weapons establishment projected a heavier warhead, the R-7 was much bigger than Atlas, with a core stage plus four strap-on boosters. It would prove to be an impractical weapon but a wonderful launch vehicle, with a lifting power no American rocket could match in the early years of the space race.

The acceleration of the missile race reinforced space advocates' public messages that orbiting satellites and even humans might be imminent in space. Since 1946, scientific packages and even monkeys had been thrown on brief (and often fatal) journeys into space from White Sands and other locations. A JPL WAC Corporal sounding rocket mounted on a V-2 soared 250 miles from the Earth in 1949. The Soviets followed secretly in the 1950s, even sending dogs on suborbital trips. Space advocates like Arthur C. Clarke of the British Interplanetary Society and Willy Ley, who became a successful American science writer during and after World War II, published influential new books. In 1952, von Braun finally made a breakthrough in a series of *Collier's* magazine articles outlining his grand (or grandiose) visions for a space station and human expeditions to the Moon and Mars. Three broadcasts about space with Ley followed on Walt Disney's national television program. On the Soviet side, in a very different environment, Korolev and other interwar space enthusiasts worked to legitimize spaceflight with the political and military establishment by tying it to another revival of the reputation of Tsiolkovsky, who had died in 1935. They gave speeches, held meetings, and wrote popular articles, often under assumed names because of the deep secrecy in which they worked. On both sides of the "Iron Curtain," science fiction novels and films reinforced the message. By 1955 spaceflight indeed seemed to many ordinary people to be just around the corner.[24]

Conclusions

Advocacy by true believers was critical for convincing ordinary people and elites that space travel was not a crazy idea. But it is impossible to imagine its remarkably accelerated arrival without World War II and the Cold War. The Germans' ill-advised decision to develop the V-2, followed by the rapid collapse of the alliance that defeated Hitler's Reich, accelerated the coming of the space booster by at least a decade. Now a new proximate cause was needed to get governments to fund satellites—and once again it would be the Cold War.

2

THE COLD WAR SPACE RACE

When the Soviet Union launched Sputnik ("fellow traveler" or "satellite") on October 4, 1957, it was a landmark moment. For the first time in history, humans accelerated an object to over 17,000 mph and put it into Earth orbit. The world press immediately heralded the coming of the "Space Age," and quickly labeled the ensuing superpower competition the "space race." It was a competition that would propel space travelers to the Moon in less than twelve years. But the race really began in summer 1955, when both sides announced that they would orbit scientific vehicles for the International Geophysical Year (IGY) in 1957–1958.

The First Satellites

In the United States, the IGY satellite's origins go back to a June 1954 meeting at the Office of Naval Research.

For the first time in history, humans accelerated an object to over 17,000 mph and put it into Earth orbit.

Among the attendees was Frederick C. Durant III, a rocket expert, navy reserve officer, and covert Central Intelligence Agency (CIA) officer in the scientific intelligence unit. He was also president of the International Astronautical Federation (IAF), formed in 1951 to link space societies, still mostly European, in a new transnational network. He invited Wernher von Braun to join the satellite meeting—the two had become friends when Durant read the German's paper at a European IAF congress.

Von Braun came with a low-cost launch vehicle proposal: three stages of small, clustered anti-aircraft rockets on top of his Redstone missile. Its capability was minimal: subsequent studies would only guarantee orbiting an inert sphere of 5 pounds. Von Braun and his associates argued that it was a quick and cheap way to beat the Soviet Union. Durant got the CIA to endorse the proposal because of the satellite's potential impact on global opinion in the Cold War. The question became how to track it optically to extract scientific information about Earth's extreme outer atmosphere and gravitational field—knowledge that was critical to improving the accuracy of long-range missiles. In January 1955, with joint army and navy support, the secret project was officially dubbed Orbiter.[1]

Scientists, engineers, and policy experts began building support for a satellite project in public and in the classified world. The American Rocket Society, which had remade itself as a national engineering organization after

World War II, carried out a public study. Lloyd Berkner, an influential American scientific leader with close connections to the U.S. Defense Department, crafted a resolution at the world geophysical congress in fall 1954 that advocated launching satellites for the IGY. That scientific campaign, which was to run from July 1, 1957, to December 31, 1958, as the Sun's activity peaked, particularly targeted Earth's polar regions, atmosphere, and ionosphere. Satellite measurements would yield much new data, but Berkner was also motivated by Cold War concerns, including U.S. national prestige and, very likely, the precedent that a satellite would set for overflying other countries.[2]

During that same year, 1954, the Eisenhower administration was conducting a deeply classified study of the threat of surprise attack by Soviet nuclear-armed bombers and missiles. The study endorsed several measures, including the intermediate-range ballistic missile as a stopgap, an ultra-high-altitude reconnaissance aircraft to overfly the USSR illegally (soon called the U-2), and a satellite to set the precedent for space-based reconnaissance, the ultimate solution to the intelligence problem. American experts in international law had argued that national control of air space stopped at the sensible atmosphere and therefore an orbiting object could operate freely. In spring 1955, that argument was one of the primary reasons why President Dwight D. Eisenhower decided to endorse an open, scientific IGY satellite project with a hidden policy agenda.[3]

In the Soviet Union, Sergei Korolev and his associates, notably aerospace engineer Mikhail Tikhonravov, had tracked the American and European spaceflight literature and had worked to build enthusiasm at home. Between 1953 and 1955, Tikhonravov led a small team at his institute in writing a long report on the potential of satellites, paralleling similar studies by the Rand Corporation, a U.S. Air Force–funded think tank. Military applications were featured, in part because it would appeal to Soviet decision makers. But reconnaissance, although noted, was not central to Soviet concerns because it was much easier to spy on the relatively open United States. In late 1954, Korolev, Tikhonravov, and other advocates got the prestigious Academy of Sciences to form a spaceflight commission. When a Moscow newspaper announced its existence in April 1955, there were unintended consequences. Western media picked up on the article, trumpeting evidence that a space race was coming. That reinforced the Eisenhower administration's decision to have a satellite project, although not one that would interfere with ICBM development.[4]

Orbiter, meanwhile, had picked up momentum as team members studied ways to improve its lifting capability and tracking. But its seemingly inevitable coronation as the official U.S. project was derailed in summer 1955. Milton Rosen of the Naval Research Laboratory (NRL), chief engineer of the Viking sounding rocket, proposed a new three-stage vehicle based on scaling up Viking. It quickly ended navy support for Orbiter. The U.S. Defense Department created a selection panel to choose among those two and a long-shot air force proposal based on the Atlas ICBM. The latter was expensive, would take longer, and lacked high-level support in its own service. Gen. Bernhard Schriever thought it was a distraction from his ICBM and reconnaissance satellite programs. In early August 1955, the committee surprisingly picked the NRL proposal by a vote of five to two. Vanguard offered a payload of at least 20 pounds, which when combined with a radio transmitter, promised a much larger scientific return. A secondary consideration may have been that the rocket looked more "civilian" because it used no military missile as a stage. One committee member also thought that the German origin of von Braun's rocket was a disadvantage; he may have harbored resentment against the ex-Nazi engineer. But these secondary reasons were not determinative, and the overflight argument played no role; indeed it appears to have been too deeply classified for most of the committee to know about it. The army and von Braun were stunned that a rocket based on available hardware lost out to one that needed a lot of development. There was a last-minute attempt at a compromise, wherein the army would launch the navy's satellite, but it lost. When the story of the Vanguard decision came out two years later, after Sputnik, it unleashed a lot of recriminations in the United States.[5]

Days before the committee reached its initial decision, the White House had announced the IGY satellite project on July 29, 1955, making world headlines. In Copenhagen, Denmark, the IAF was holding its annual congress, the first that Soviet delegates attended. On August 2, Leonid Sedov, the nominal head of the spaceflight commission, told the press that the Soviet Union would launch satellites too. This statement seems to have lacked top-level political backing, but may have been orchestrated by Korolev and others in Moscow, who worked in state-imposed secrecy. In 1955–1956, they formulated and got approval for a very large, half-ton geophysical satellite that would eventually be launched as Sputnik 3 in 1958, thanks to the lifting power of the R-7. But in late 1956, Korolev became concerned that the Soviets might come in second to the United States in the race for the first satellite. At Redstone Arsenal, von Braun and his army superiors had kept their project alive, convinced that Vanguard would fail. Helped by Eisenhower's decision to approve the Jupiter intermediate-range missile, they formulated a plan to use an improved version of the Orbiter launch vehicle to test heat shield technologies so that nuclear warheads would survive reentry. It would be called Jupiter-C to capture the priorities of that program, although the first stage was a Redstone. A September 1956 test produced a world record distance shot of 3,355 miles. Korolev, based on Western press rumors, was convinced von Braun had made a failed satellite attempt. His group formulated a plan for a simpler, radio-equipped sphere—the later Sputnik 1. It could be launched once the first R-7 ICBM tests succeeded.[6]

Just as von Braun (now a U.S. citizen) had expected, NRL's Vanguard ran increasingly behind schedule and over budget, although Rosen and his compatriots would eventually master the development problems. The army twice attempted to get Eisenhower to bless Jupiter-C as a backup to Vanguard, primarily to ensure that the United States would be first. But the president was not interested. It would cost more money at a time when Vanguard budgets had already gone up dramatically, and it would undercut U.S. policy that Vanguard was the official scientific IGY project. These decisions, like the original choice of Vanguard (in which Eisenhower had no role, other than not blocking it) were fundamental in their consequences: the history of the space race would have been completely different if the United States had gone first. There would have been no need to catch up. Once again, the course of events accelerated the arrival of spaceflight capability almost to the maximum conceivable extent.

Organizing Space Programs

Sputnik's impact on American public opinion is often exaggerated. Popular historians have used words like "panic,"

Sputnik's impact on American public opinion is often exaggerated. Popular historians have used words like "panic," "hysteria," and "fear" to describe its effect.

"hysteria," and "fear" to describe its effect. Recent studies do not bear this out. Many Americans were impressed by the achievement or were just indifferent. But the press and politicians quickly jumped on the Eisenhower administration for having given the Soviets this symbolic victory.[7] It fed into an existing media narrative that the president was a kindly old man who would rather play golf than govern. "Ike" preferred to keep his deep involvement in Cold War decision making hidden, in part to protect his secret agendas, such as the overflight strategy and the dangerous U-2 reconnaissance missions. Those flights, limited as they were, suggested that far from there being a "bomber gap," and later a "missile gap," the inverse was true: the United States was ahead in every category of nuclear arms. But his critics, lacking that knowledge, had a field day criticizing the president's unwillingness to feed the arms and space races further. The critics' agendas were often in conflict, however. The army and air force were bitter rivals over ballistic missile development, a rivalry that instantly spread to the space race. Both staked claims to run the future program, while the navy argued for its own piece of the action.

The global press reaction also caught Soviet leaders by surprise. The official newspapers carried a small article on the day after launch, but on October 6 there were huge headlines as international congratulations poured in. Communist Party head Nikita Khrushchev, who had consolidated power after Stalin's death in 1953, quickly

The army and air force were bitter rivals over ballistic missile development, a rivalry that instantly spread to the space race.

wanted another space spectacular for the fortieth anniversary of the Bolshevik Revolution in early November. Korolev's team threw together a dog mission, using cabin equipment developed for suborbital flights. On November 3, Sputnik 2 orbited with Laika, a stray dog picked off the Moscow streets. Unfortunately the poor dog panicked and then died due to overheating, but the Soviets lied about it for a week before claiming she had been humanely euthanized. The large size of the spacecraft, 1121 pounds, many times the 184 pounds of Sputnik 1, was quite impressive and legitimized again the Soviets' claim that they'd made the first successful ICBM test in late August.[8]

The second success exacerbated what historian Walter McDougall has called "a media riot" in America over the perceived embarrassment. U.S. public opinion slowly tilted toward anger, worry, and criticism of the administration. The Soviet ICBM threat, although in fact years away from being effective, heightened the sense of vulnerability. Khrushchev was happy to feed this fear by boasting about Soviet missile and space capabilities. To make matters worse, on December 6, a Vanguard rocket carrying a mini-satellite fell back on the launch pad, exploding spectacularly on national TV. By this point, the army had succeeded in getting approval for its backup project based on Jupiter-C. The Redstone missile and project direction came from von Braun's team in Huntsville, while the solid-propellant upper stages and satellite were managed by the

Jet Propulsion Laboratory in California. On January 31, 1958, the United States at last had a satellite in orbit; the army named it Explorer I. Vanguard followed with its first orbital success six weeks later.[9]

As the military services fought over the space program, both the Republican president and the Democratic senate majority leader, Lyndon Johnson, came to accept the need for a civilian agency to carry out peaceful and scientific missions. The most obvious foundation was the National Advisory Committee for Aeronautics (NACA), a government research organization founded in 1915, with major centers in Virginia, Ohio, and California. At the end of July, the President signed the bill to create the National Aeronautics and Space Administration (NASA) out of NACA, plus the Vanguard group at NRL and some army and air force space projects. Academically run JPL, which was managed by Caltech, was happy to defect from the army to NASA soon after the agency began business on October 1, 1958. But von Braun and his commander in Huntsville fought off an attempt to transfer half of his group that fall, fearing the effects of a breakup on the Jupiter missile and other projects. Eisenhower finally ordered the transfer a year later, when all of von Braun's personnel could be accommodated.[10]

Creating a civilian agency was an American Cold War solution to the organizational, but also the political, challenges of the rapidly escalating race. The Sputniks had

Both the Republican president and the Democratic senate majority leader, Lyndon Johnson, came to accept the need for a civilian agency to carry out peaceful and scientific missions.

demonstrated the value of space achievements for garnering prestige and signaling scientific and technological strength. The breakup of the European colonial empires, notably those of Britain and France in Africa and Asia, was an important context. New nations were forming every year, and nationalist revolutionary movements often looked to the USSR and China as models for development. Soviet space accomplishments were a powerful advertisement for the alleged superiority of socialism over capitalism. Communist propaganda relentlessly criticized America and the West for militarism and imperialism. A civilian and scientific space agency, with international cooperation written into its founding legislation, would project a positive American image to allies in Western Europe and elsewhere. By creating NASA, the United States also effectively invented the category of civilian space activity, as heretofore only the armed services possessed the technologies that made spaceflight possible.[11]

The creation of NASA conveniently misled many to think that it ran "the" U.S. space program. In the first four years after Sputnik, the American government actually created three such programs. The first was the NASA-controlled civilian effort. The second was the U.S. Air Force–dominated military one, the army and navy having lost much of their space personnel to NASA. Reconnaissance, communications, navigation, and other military satellite systems originated in this period, and the USAF

Soviet space accomplishments were a powerful advertisement for the alleged superiority of socialism over capitalism.

also pursued its dream of piloted space planes, even as NASA was handed the mission of putting someone into orbit. The third U.S. space program was the intelligence one, closely tied to, but organizationally distinct from, the military program. It began with the super-secret CORONA film-return reconnaissance satellite project, which Eisenhower split away in early 1958. Like its parallel black program, the U-2, it would be run as an air force-CIA collaboration. The navy then contributed the first signals intelligence satellites. In 1961, the multi-agency program was formalized as the National Reconnaissance Office (NRO), the very name of which was secret until 1992. It built and operated spy satellites, in close collaboration with the air force, but handed the products over to the intelligence agencies. NASA also worked closely with the CIA and the military services, particularly on intelligence about the Soviet space program, sensor technology, and launch vehicles, but hid that much of that collaboration behind a wall of classification to protect its image as peaceful.[12]

The Soviets never felt the need to create a civilian agency even as a front. Cloaked by secrecy, their program was effectively military. Khrushchev split the ballistic missile troops away from the army and created a separate service, the Strategic Rocket Forces, which carried out all space launches. Military-oriented factories, institutes, and design bureaus, like Korolev's OKB-1, built the technology. The air force selected and trained the first cosmonaut class in 1960. To the outside world, the USSR Academy of Sciences represented the program; in reality it was only involved in scientific experiments, although the prestige of leading academicians could be valuable in internal Soviet politics.

The early space race was punctuated by Soviet spectaculars that overshadowed America's advantage in other sectors. In 1959, Luna 1 flew by the Moon and became the first object to escape Earth's influence, Luna 2 hit the Moon, and Luna 3 made crude photos of the side never seen from the home planet. The best the United States could do was fly by at a much greater distance. On April 12, 1961, Yuri Gagarin became the first human in space, making one orbit in Vostok 1; four months later, Gherman Titov circled for an entire day. The American Mercury program struggled with delays. Alan Shepard and Virgil "Gus" Grissom made two short suborbital hops between the Russian flights. It was not until John Glenn orbited in February 1962 that U.S. efforts equaled the Gagarin achievement. In June 1963, the Soviets orbited the first woman, Valentina Tereshkova, while NASA resisted attempts to add women astronauts.[13] Behind the scenes, however, the United States was two years ahead in reconnaissance satellites, having had its first successes with navy signals intelligence payloads and CORONA photo reconnaissance satellites in spring and summer 1960. In science, the United States was also ahead, with more launches

In June 1963, the Soviets orbited the first woman, Valentina Tereshkova, while NASA resisted attempts to add women astronauts.

and more effective payloads. But one would hardly know it from the press reaction after each Soviet triumph.

President Eisenhower resisted increasing the national debt and the size of government, but missile and space programs grew anyway because of public and political pressure to match the Soviets in the arms and space races. The former five-star general famously left office denouncing the "military-industrial complex" and the elite experts who tried to sell increasingly expensive projects—much as von Braun did. In contrast, John F. Kennedy came into office in January 1961 in part by running on the "missile gap" and America's space inadequacies.

The Moon Race

Two crises in April forced Kennedy's hand: the Gagarin flight and the ignominious failure of a Cuban invasion by CIA-supported exiles five days later. He tasked his vice president, Lyndon Johnson, to find an aspect of the space race "we could win." NASA already had its answer: "landing a man on the Moon and returning safely to the Earth," as Kennedy put it in his May 25 speech to Congress. Anything less than a full human lunar landing, NASA's new administrator, James Webb, estimated, would not insure that the United States had a good chance of beating the Soviets. The rocket required would be so large that it

Figure 3 Yuri Gagarin, the first human in space, is hailed by Soviet leader Nikita Khrushchev in Moscow's Red Square after his historic one-orbit flight on April 12, 1961. A string of Soviet firsts powered the early U.S.-Soviet space race and motivated President Kennedy to propose sending astronauts to the Moon by the end of the 1960s. *Source:* Smithsonian National Air and Space Museum (NASM 82-8652)

would effectively cancel out their current advantage in lifting power. When Shepard's flight went well on May 5, it provided further momentum for the decision. The estimated budget was $20 billion to $40 billion, an enormous price by the standards of the day. The timetable was also mind-boggling: "before this decade is out" was how Kennedy put it, meaning 1969 or maybe 1970. Yet Congress was so much in agreement that it rubber-stamped big increases in the NASA budget. Once again, international rivalry and the accidents of history combined to accelerate the timetable, such that humans would step on the Moon only eight years after the brief Gagarin and Shepard flights.[14]

The Moon goal transformed NASA. By 1966, the agency's expenditures quintupled to $5 billion. It funded a boom in the aerospace industry, notably in California, and paid for massive new facilities across the southern United States. These included a Manned Spacecraft Center (later the Johnson Space Center) in Houston, Texas, plus massive expansions of von Braun's Marshall Space Flight Center in Alabama, and its spin-offs, the Florida launch center named for Kennedy after his 1963 assassination, and a rocket testing facility in Mississippi (now Stennis). Human spaceflight became NASA's dominant mission. When the Apollo boom came to a surprising end in the late 1960s and early 1970s, that infrastructure and budget dependence would be a problem.

But that seemed far in the future. In 1961–1962, the agency picked a launch vehicle, the Saturn V, and a landing method, lunar orbit rendezvous, meaning that it needed to develop a separate, specialized lunar lander for Apollo, in addition to the main spacecraft. The Houston center, headed by Mercury program chief Robert Gilruth, also decided that it needed a bridge from its first human spacecraft to Apollo: a scaled-up, two-man Mercury called Gemini. Its primary objectives would be to gain experience with rendezvous and docking, spacewalking, and the medical effects of spaceflights up to fourteen days, knowledge needed to make the challenging lunar journey. Despite a few crises and problems, NASA carried out ten crewed Gemini missions in twenty months in 1965–1966, accomplishing all objectives. That was the period when the United States caught up and passed the Soviet Union in the Moon race.[15]

This lead was not immediately obvious as Korolev's team pulled off a few more firsts. In October 1964, the Soviets launched three cosmonauts in Voskhod 1, and in March 1965, two in Voskhod 2. One of the latter, Alexei Leonov, became the first person to walk in space. Carefully concealed from the world was how dangerous these missions were. To keep producing spectaculars for Khrushchev, Korolev had modified Vostok by taking out the ejection seat and cramming in more cosmonauts. There would be no escape if the booster were to fail. Leonov faced a serious crisis during his spacewalk, when his suit ballooned so much that was difficult for him to get back into Voskhod 2's inflatable airlock. The spacecraft also had control problems and landed far off course. Leonid Brezhnev, Alexei Kosygin, and other party leaders had in fact forced Khrushchev to resign a day or two after Voskhod 1. Leonov's March 1965 walk was the last stunt of the old style. A puzzling gap followed, in which no cosmonaut was launched for two years. Korolev's design bureau struggled behind the scenes to perfect its new spacecraft, Soyuz, and its Moon-bound derivatives.[16]

Despite two more firsts with robotic Moon missions in 1966, the mid-1960s were the period when the Soviet space program began to fall apart. The reasons were many. During the ballistic missile drive in the late 1940s and 1950s, Korolev had been brilliant in leading and coordinating various design bureaus and industries. By the 1960s, the maturation of the rocket and space industry meant that there were now several large enterprises led by important personalities who competed—often bitterly—for favors from the military-industrial complex and party leadership. Korolev's R-7 was an impractical ICBM, so Mikhail Yangel's design bureau won with better designs. Korolev had a falling out with Valentin Glushko, the primary liquid-fuel rocket engine designer, over the selection of propellants for the giant N-1 booster that was to be used for the lunar landing, and gave the engine design to another enterprise.

Rivalry also erupted with the designer Vladimir Chelomey, who developed the Proton booster and alternate plans for human missions, including a project to loop two cosmonauts around the Moon. The latter program was eventually transferred to Korolev's bureau, creating two lunar projects that spread scarce resources even thinner.

As it was, the Soviet decision to compete with Apollo came shockingly late. Khrushchev only approved a program in August 1964. The string of Soviet successes seems to have made everyone complacent. In early 1966, Sergei Korolev died in a botched medical operation, robbing the space program of critical leadership. But the increasing conflict demonstrates that the one-party dictatorship and planned economy were less successful that the democratic capitalist system in mastering internal economic competition and creating a coherent program—contrary to contemporary Western expectations that dictatorships were more likely to make clear decisions. On top of all this, the Soviet Union simply did not have an economy that was large enough and efficient enough to support a Moon race and an all-out missile competition with a mobilized United States. Brezhnev and Kosygin underfunded the lunar projects and prioritized catching up with the United States in nuclear delivery systems. The Cuban Missile Crisis in October 1962 had been humiliating; Khrushchev had to remove rockets put there to compensate for Soviet strategic inferiority.

In early 1967, both superpowers' human programs had tragic setbacks that again obscured the growing American lead. In January, the first Apollo spacecraft that was to carry a crew killed its three astronauts—Gus Grissom, Edward White (who had made the first American spacewalk in 1965) and Roger Chaffee—in a launch-pad fire. That plunged the Apollo program into crisis, as it revealed how flawed the spacecraft was. Three months later, in April, Vladimir Komarov died in the crash of Soyuz 1, when his parachutes became entangled at the end of a crisis-ridden flight. Neither country flew crews again until late 1968.

For NASA, the fire forced a thorough overhaul of the Apollo program, resulting in a stunning string of successes between late 1967 and late 1969. The perfect first test of the gigantic Saturn V rocket was followed by an orbital test of the lunar module, then five straight astronaut missions starting in October 1968. Notably, Frank Borman, James Lovell, and William Anders made the historic first human journey into deep space in Apollo 8, orbiting the Moon at Christmas. Two missions followed in Earth and lunar orbit now that both the mothership and the lunar lander were available. The culminating triumph was Apollo 11, when Neil Armstrong and Buzz Aldrin landed and walked on the Moon on July 20, 1969, while Michael Collins orbited overhead. Their safe return, with the first samples from another heavenly body, marked the achievement of Kennedy's 1961 challenge. Late in the year, Apollo 12 followed with a precision landing near Surveyor 3, one of America's robotic explorers that had been there since 1967.

The Soviets carried out several Soyuz missions in low Earth orbit over the same period, including docking two of them. But these missions, beyond developing space experience, mostly served to cover the complete failure of their lunar programs. A stripped-down Soyuz equipped for circumlunar flight was to carry two cosmonauts around the Moon before the Americans. Partly successful unmanned tests made it appear as if the USSR were in a close race with Apollo 8. But no crew ever launched because fixes were needed, and then the United States struck first and more impressively, by orbiting a crew, not just looping them around the Moon. The first two tests of the Saturn V-sized N-1, which would launch the lunar landing spacecraft, were disastrous failures in 1969, thanks to an overly complex first-stage design and minimal testing, byproducts of a lack of money. The N-1 program dragged on into the early 1970s, but two more launch attempts ended as badly as the first two. In public, the Soviet Union denied it ever had a human lunar landing program.

The Space Race Shifts into a Lower Gear

The Apollo 11 triumph ended the first phase of the space race. NASA was already taking budget cuts as Apollo-Saturn spending peaked in 1966 and Vietnam, urban riots, and other national problems undercut public support. The Moon victory inspired NASA to try to get the new Nixon administration to approve a space shuttle, space station, lunar exploration, and a human Mars expedition by the 1980s. But the country was in no mood for that, and neither was Nixon. NASA's budget kept falling until the mid-1970s, when it effectively had half the buying power it did in 1966. On the Soviet side, the secret failures of the lunar program refocused Korolev's heirs on a series of small orbital stations. Eventually, the Soviet leadership also decided to build a space shuttle, in imitation of the only substantive new human spaceflight program that NASA got in the early 1970s. The American one was sold to Nixon and the Congress on the argument that a reusable vehicle would drastically lower the cost of launch. More ambitious plans would be postponed until the 1980s or later.[17]

Thus the forces of war and international competition, which had repeatedly accelerated rocket and space technology development, suddenly slackened. The United States carried out four more Apollo landings in five attempts, but when that program ended in December 1972, it marked the last time humans would venture more than 400 miles from Earth, at least as of this writing. In hindsight it seems apparent that such a lopsided historical trend was unlikely to continue. But it came as an unpleasant surprise to the true believers, who had expected that human deep-space exploration would go on forever.

Yet the space race was not over; competition went on for another twenty years in moderated form. As long as the Cold War continued, neither the United States nor the USSR could break the cycle of trying to match the other's accomplishments. This was notably true in the military space realm, in which the competition for space capabilities continued unabated, even during the mid-1970s period of détente, when the two sides arranged a public show of cooperation by docking an Apollo and a Soyuz spacecraft in 1975. Both superpowers continued to develop their photo and signals reconnaissance capabilities, as well as early-warning, weather, communications, navigation, and other satellite systems for military and national security purposes. The Soviets even tested anti-satellite weapons and orbital bombardment systems to attack the United States.[18]

Despite this intense and often hidden competition, both sides had come to accept a de facto regime where space was militarized but not weaponized. After a tense period in the early sixties that included space nuclear tests, the U.S. and USSR agreed to a Partial Test Ban Treaty in 1963 and a UN Outer Space Treaty in 1967 that forbade orbiting "weapons of mass destruction." Both powers kept only a minimal capability to attack each other's spacecraft from the ground or from co-orbiting satellites. It simply was not in either superpower's self-interest to pursue a space arms race; they had become dependent on their respective satellite networks.

Another important dynamic of the later Cold War was that space activity became increasingly less bipolar. This trend began in the early 1960s, when Britain and Canada developed scientific satellites to be launched by the United States, followed in 1965 by France becoming the third nation to put its own satellite into orbit. Feeling technologically inferior to the United States, Western European nations also created two cooperative agencies, one for scientific satellites and one to develop a civilian launcher based on British and French ballistic missiles. The former organization was a success and the latter a miserable failure thanks to poor systems management across national boundaries. In 1975, the two organizations merged into the European Space Agency (ESA), which was founded on a German-France compromise: the Germans were more interested in cooperating with the United States on human spaceflight programs, while the French again wanted to try to develop an independent launch vehicle capability. That led to the success of the French-dominated Ariane rocket, which seized a remarkable market share in the 1980s by launching communications satellites into 24-hour orbits. In Asia, Communist China orbited its first satellite in 1970, coming out of a ballistic missile program headed by rocket engineer Qian Xuesen, who had spent many years in the United States before being forced to leave on false charges of disloyalty. Japan also launched its first satellite in 1970. Its satellites and boosters came partly through indigenous effort and partly by cooperating with the United States. In 1975 the Soviets launched India's first satellite, but five years later the Indian Space Research Organization succeeded in orbiting one on its own.[19]

In human spaceflight, the 1970s and 1980s saw the development of the first orbital stations. Coming out of a much-reduced program for exploiting Apollo-Saturn technology, NASA astronauts occupied a Skylab space station in 1973–1974 based on a Saturn V stage, but there was no money for a follow-up. The Soviet Union flew two different types of small stations in the 1970s and 1980s, but called them both Salyut to obscure the fact that one was a military reconnaissance spacecraft for spying on the United States and its allies that successfully flew twice. That project was a response to the USAF-led Manned Orbiting Laboratory, which never got off the ground because the Nixon administration decided that the job could be done adequately by unmanned satellites. But the Soviets did accumulate extensive experience with the long-duration effects of weightlessness on its Salyut stations, and their follow-on, Mir, launched in 1986.[20]

Not surprisingly, when NASA set out to sell a space station in the early 1980s, after finally flying the U.S. Space Shuttle, it used the specter of Soviet superiority in advocacy to the conservative Reagan administration. However, it appears that the space agency's primary motivation had more to do with getting back to what one political scientist has called the "von Braun paradigm"—space shuttle, space station, Moon, and Mars—as the "logical" stepping-stones of human spaceflight. Because Europe, Canada, and Japan now had capability developed in working with the U.S. shuttle, and such cooperation promised to save American taxpayers money, the space station program was international from the outset. But it took until the end of the Cold War before anyone could contemplate a post-Soviet Russia as a partner, and it took until 1998 before the International Space Station (ISS) began assembly in orbit.[21]

Instead it was the shuttle that dominated NASA's program in the 1970s and 1980s. What had once been thought of as just a space station transport vehicle became an end in itself after the budget cuts of the early 1970s. To save it, the space agency made a deal with the U.S. Defense Department to make the shuttle the standard launch vehicle for all programs. The wing shape and the payload bay size were determined by the requirements of secret military missions. The air force built a separate California launch pad, never used, for polar orbit launches. In order to reduce peak development cost, NASA decided on a partially reusable system, with recoverable solid-rocket boosters but a disposable propellant tank. Wildly optimistic estimates for how often the shuttle could fly (almost once a week) and how much it would cost to put a pound in orbit ($100—it actually turned out to be more like $10,000) were made to sell the program to Congress and Nixon. In the late 1970s, the technologically challenging orbiter project ran into major problems with its reusable reentry tiles and its main rocket engine, delaying launch by over two years. John Young and Robert Crippen finally flew *Columbia* into orbit on April 12, 1981, the twentieth anniversary of Gagarin's historic flight.[22]

The shuttle was definitely a mixed bag for the United States. On the one hand, it was a technological marvel, the world's first (mostly) reusable human spacecraft. It enabled expanded access to low Earth orbit by non-pilots, including the first female and minority astronauts (Sally Ride and Guy Bluford) in 1983. The first satellite repair and recovery missions took place in the early 1980s, paving the way for salvaging and maintaining the Hubble Space Telescope (HST) in the 1990s and 2000s. Shuttles carried European-built laboratories and instruments and a Canadian-built manipulator arm into orbit, along with astronauts from those countries and others. On the other hand, the shuttle was an economic failure, as it cost far more time and money to refurbish the orbiter before each flight than was anticipated, and it was dangerous. The first shuttle accident, the explosion of the *Challenger* on January 28, 1986, killed seven astronauts and stopped all flights for two-and-a-half years. It confirmed for the military and NRO what it had already concluded, that the shuttle was not a reliable launcher for national security missions. Afterward, President Reagan ended the "all eggs in one basket" policy of trying to replace all other rockets, and he stopped NASA sales of commercial satellite launches (which had massively lost money, a fact the agency could not obscure). The shuttle gave the United States a lot of international prestige, but it was also a drain, leaving human spaceflight stuck in low orbit, even as American robotic spacecraft spectacularly flew across the solar system. The shuttle sustained the human spaceflight infrastructure that NASA had built for Apollo, but sometimes that seemed as if that was its main purpose, at least for the agency bureaucracy and for congressmen with space centers or contractors in their districts.[23]

In the mid-to-late 1980s, the military space competition between the United States and the USSR seemed to build to a new dangerous level, and then just as suddenly faded away. When Reagan made his "Star Wars" speech in March 1983, calling for a ground-and-space-based missile defense against nuclear attack, it threatened to destroy the de facto regime of a militarized, but non-weaponized, near-Earth space. It caused an international uproar and empowered Soviet military space organizations to ask for more money for laser battle stations, anti-satellite weapons, and missile penetration aids. It seems exaggerated to

Figure 4 Space shuttle *Discovery* launches on its first mission, August 30, 1984. The shuttle became the primary human spaceflight program after the Moon landing and dominated U.S. space policy in the late and post–Cold War eras. It was a technical success but failed to revolutionize the cost of spaceflight. *Source:* NASA

assert that Reagan's defense buildup and Strategic Defense Initiative (SDI) pushed the Soviet Union into collapse, especially since the stagnation and dysfunction of the Stalinist planned economy had deep roots. But SDI made an impression on party leaders like Mikhail Gorbachev, who rose to the top in 1985, and empowered his push for arms control agreements so that money could be diverted to the ailing civilian economy.[24] The result was the sudden de-escalation of the nuclear arms race, shockingly followed by the 1989 collapse of the Soviet empire in Eastern Europe and the disappearance of the USSR itself in 1991. With it went the bipolar space race, and a significant motivation for space spending.

Conclusions

The Cold War space race was the fundamental shaping influence in the history of spaceflight. Driven by a super-power competition to signal technological capability and military strength, the United States, the Soviet Union, and other powers hurriedly built government organizations, manufacturing capacity, and operational experience sufficient to support a broad spectrum of space activity, notably in human spaceflight. The race also stimulated the scientific exploration of the universe and a growing, global infrastructure in orbit. These would prove self-sustaining even when the Cold War itself disappeared.

3

SPACE SCIENCE AND EXPLORATION

In the decades since 1945, knowledge of Earth, the solar system, and the universe has been transformed by the accelerated achievement of spaceflight, with significant impacts on the understanding of our origin and place in the cosmos. What began as a tentative, rocket-borne examination of the upper atmosphere and near-space, primarily to aid Cold War military operations, led quickly to the space race–driven exploration of the Moon and the planets, and to the creation of space telescopes able to observe at wavelengths blocked by the protective power of our atmosphere.

The space sciences greatly benefited from an unforeseen robotic spacecraft revolution. To early advocates, humans seemed essential to navigate the spacecraft, point the instruments, and take the observations, and in any case, sending people into space was just assumed to be the point. But the post–World War II miniaturization of electronics and instruments capable of withstanding the vibration, shock, temperature extremes, and vacuum conditions of launching and spaceflight made remotely operated spacecraft not only feasible, but much cheaper than human spaceflight, which required elaborate technologies to protect and sustain life. Nonetheless, astronauts and cosmonauts sometimes actively carried out space science missions, whether to satisfy political considerations, like the Moon race, or because studying the adaptation of human bodies to spaceflight was the science.[1]

While the Cold War was the primary driver of space science and exploration before 1989, these activities remained surprisingly vigorous afterward, except in post–Soviet Russia. U.S. space science became a self-sustaining enterprise because it created institutions, industries, and capabilities that politicians saw as valuable to national prestige, defense capability, the domestic economy, or their home districts. Similarly motivated, Western European and Asian governments began expanding their space science programs in the 1960s, eventually filling the gap left by the dramatic decline in Russian capability.

Reaching into Space

While Robert Goddard and the Peenemünde group had made plans to launch instruments into the upper atmosphere, it fell to American university and government laboratories to begin experimenting immediately after World War II. Resources at first were thin as the country tried to demobilize. A series of ad hoc experiments began at White Sands, New Mexico, mostly using captured V-2s prepared by military and General Electric personnel, assisted by von Braun's German team. Money was lacking to create parachute recovery systems, such as the Germans had planned, so scientists from the Naval Research Laboratory, the Johns Hopkins University Applied Physics Laboratory (APL), and others cobbled together instruments that could survive impact in the desert or radio back some data. Failures were so numerous that scientists who were primarily interested in answering questions about the upper atmosphere, space, or the Sun mostly quit, leaving experimentalists from science and engineering who were fascinated with building something that worked on a rocket.[2]

The U.S. armed services, and soon those of the USSR, Britain, Canada, and other countries, facilitated rocket experiments because they wanted to understand the environment in which high-speed aircraft and guided missiles would operate, and because it would shed light on how the ionosphere—layers of charged particles in the extreme upper atmosphere—reacted to solar activity. The ionosphere reflects longer-wavelength radio waves, which had become critical to communication and defense. The Cold

War arms race intensified interest in the polar regions, notably because geography dictated that a US-Soviet nuclear war would in part be conducted over the North Pole. Even before the American supply of V-2s ran out in the early 1950s, NRL, APL, and other organizations worked to develop cheaper sounding rockets like Viking and Aerobee. With the intensification of the Cold War, rocket experimentation was extended to ships at sea and the polar regions, notably during the International Geophysical Year in 1957/58, which in turn led to satellites that could put scientific instruments into space for long periods, rather than just for a few minutes.

Among the early V-2 scientists was the physicist James Van Allen, then working at APL and later at the University of Iowa. His postwar career illustrates the origins and evolution of the space sciences. He pursued his interest in lifting instruments to detect "cosmic rays"—high-speed atomic nuclei from space first detected in 1911—from the V-2 to military-funded small sounding rockets to the first Explorer and Vanguard satellites. He helped create the Aerobee sounding rocket, which evolved from Jet Propulsion Laboratory's WAC Corporal first launched in late 1945. A 1950 dinner party at Van Allen's house in Silver Spring, Maryland, sparked the creation of the International Geophysical Year, timed for the maximum in solar activity that happens every eleven years. His cosmic ray experiments on Explorers I and III in 1958 detected protons and electrons from the Sun trapped by Earth's magnetic field. These formed two radiation-intensive zones soon called the Van Allen Belts. He went on to be principal investigator of particles and fields experiments on many spacecraft sent to Earth orbit, interplanetary space, and the planets.[3]

Van Allen was not alone, although his career was especially significant. Funded primarily by the militaries of various countries before 1958, physicists, chemists, and engineers began building an institutional infrastructure, both national and international, for conducting science in or about space. In the early space race years, the ionosphere, the auroras (northern and southern lights), and the environment near Earth were the most important objects of study, with the aim of understanding the interaction of solar particles and radiation with our upper atmosphere and magnetosphere (the space region controlled by Earth's magnetic field). NASA and the USSR Academy of Sciences funded new institutes and new experimentation, and it was also the starting point for NASA's international cooperation program with Britain, Canada, Western Europe, India, and elsewhere. Out of these efforts grew the discipline of space physics, more recently called heliophysics, which acknowledges the dominant influence of the Sun on the interplanetary environment and the space near Earth.

Racing to the Moon and Planets

As I noted in chapter 2, the first two Sputniks also spun off a Moon race that led to three Soviet firsts in 1959: flyby, impact, and photographing the far side. In terms of science and exploration, the last was most important, as Luna 3's low-resolution TV images showed almost no dark, flat lava plains like those that make up much of the Earth-facing side. The competing American Pioneer programs failed as lunar missions, but three early spacecraft did send back space physics data at record distances of tens of thousands and hundreds of thousands of miles.

The two superpowers immediately jumped into a prestige race to reach the closest planets, Venus and Mars. Here the pattern of success was reversed. All nineteen Soviet spacecraft launched by 1966 fell victim to booster failures or unreliability during the months-long cruise to their targets. NASA's Jet Propulsion Laboratory, which had assumed the role of primary U.S. lunar and planetary center, had two successes out of four attempts, flying by Venus in December 1962 and Mars in July 1965. Mariner 2's lone experiment confirmed that Venus's cloud-shrouded surface was hot enough to melt lead, and thus uninhabitable. Mariner 4's twenty-one pictures revealed a less alien Mars, but one that still appeared heavily cratered, Moonlike, and uninviting. Both missions were early blows to the hope that extraterrestrial life, at least in some simple form, might quickly be discovered. Beyond the superpower race for prestige, the primary thing that excited average citizens to support planetary exploration was the possibility of life.[4]

As the human Moon race accelerated, both sides invested in missions to characterize its surface and select possible landing sites. Science took a back seat to exploring the feasibility of landing, notably in the United States, where NASA's policy frustrated many scientists. JPL's early Ranger missions carried more experiments, including a balsa-wood-wrapped ball to "hard-land" a quake-detecting seismometer on the surface. But five straight failures led to a congressional investigation of JPL and the decision to concentrate on transmitting pictures of the surface on the way to impact. The simplified spacecraft then failed again. Finally, in July 1964, the United States had its first lunar success with Ranger 7, followed by two more Rangers in early 1965. NASA also dropped most experiments from its Surveyor lunar lander, which finally succeeded on its first, much delayed touchdown in June 1966. But it was beaten to the surface by the Soviet Luna 9, which made the first successful landing in February. Both missions demonstrated that the lava plains at least had the strength to bear the weight of a large spacecraft, and they disproved one theory that the surface might be buried in feet of fine powder.[5]

That same year, the Soviet Luna 10 became the first to orbit the Moon, soon followed by NASA's Lunar Orbiter

Beyond the superpower race for prestige, the primary thing that excited average citizens to support planetary exploration was the possibility of life.

1, which began a highly successful mapping program with a camera taken from a secret spy satellite program. After the first three in that series photographed Apollo landing sites at high resolution, Lunar Orbiters 4 and 5 were put in higher polar orbits, creating a global photo map of significant value to understanding the Moon's history. Ultimately, although the Soviet Union scored the firsts, the United States collected far more data that was valuable to both human exploration and to science.

While the Apollo program had been sold to the American public as a science program as well as a prestige race, the scientific community had to be pacified because clearly science was an afterthought. In 1958, the National Academy of Sciences had formed a Space Science Board—inventing the disciplinary term in the process. Three years later, leading American scientist Lloyd Berkner had to work to bring its members along to support President Kennedy's decision to spend billions on Apollo, with the expectation that robotic science programs would also be funded. With engineers running the human program and pilots filling all the seats until the very last Moon landing, when geologist Harrison Schmitt flew on Apollo 17, it was not obvious that science would get much consideration.[6]

Yet Apollo surprisingly turned into a major contribution to lunar and solar system science, as every successful landing deployed seismometers and other instruments on the surface and brought back increasingly diverse samples

While the Apollo program had been sold to the American public as a science program as well as a prestige race, the scientific community had to be pacified because clearly science was an afterthought.

of rocks and soil—over 800 pounds by the end of the program. The last three missions carried lunar rovers, greatly increasing astronaut range, and landed at much more challenging and interesting sites. The pilots in the orbiting motherships all took photos, but the final three flights carried a dedicated science bay for orbital mapping, including radars and instruments that detected the distribution of elements on the surface. When all the samples and data were processed, they provided accurate dating of the formation of the Moon and its major features. They demonstrated that there had been waves of asteroid impacts in the early solar system, shedding light on the formation and evolution of the Earth-Moon system and other planetary bodies.[7]

The emphasis on imaging from American spacecraft and the analysis of the Apollo samples reshaped the planetary sciences in the United States by greatly expanding the role of geologists. NASA funding fueled the rise of the space sciences in general, resulting in the expansion of numerous university institutes; the U.S. Geological Survey's Astrogeology Branch in Flagstaff, Arizona; the Smithsonian Astrophysical Observatory in Cambridge, Massachusetts; JPL in Pasadena, California; and the NASA Goddard Space Flight Center in Maryland, which focused on Earth-orbiting science satellites.

In the Soviet Union, two important organizational changes in 1965 strengthened the space science community and robotic lunar and planetary program. The USSR Academy of Sciences founded an Institute of Space Research, funded by the ministry that controlled the rocket and space design bureaus and factories. It built on the work of the Vernadsky Institute, which primarily focused on geophysics and geology. That same year, Sergei Korolev transferred the robotic program to the Lavochkin design bureau, headed by Georgi Babakin, because his organization was overwhelmed with human spaceflight and ballistic missile projects. Babakin's group was able to focus on reliability, although they still had problems with launch vehicles supplied by other enterprises.[8]

The Soviet Luna achievements of 1966 were followed by a string of successes at Venus. Venera 4 successfully penetrated its atmosphere in 1967. Venera 7 in 1970 became the first spacecraft to land and transmit from the surface of another planet, having been strengthened to withstand the crushing atmospheric pressures and hellish temperatures of the Venusian surface. Later spacecraft returned images, although no landers lasted more than an hour or two because of the heat soaking in from outside. Lavochkin also had some success in the 1970s with lunar missions, returning three small soil samples and driving two rovers on the surface. The Soviets found Mars, on the other hand, to be frustrating. Numerous missions failed, and while Mars 3 was the first to land on the surface, in 1971, it died twenty seconds after landing. Because the

United States was succeeding with more sophisticated Mars spacecraft, the Soviet space program leadership decided to focus on its one area of success, Venus.[9]

On the American side, the space race powered what one program veteran called "the golden age of planetary exploration" in the 1970s and into 1980s, as a kind of echo effect of the investments made in the 1960s, even as NASA budgets fell. JPL's Mariner 9, which in 1971 became the first to orbit and globally map Mars, revealed a much more scientifically interesting landscape with giant volcanoes, canyons, and evidence of a distant past in which massive floods carved out numerous channels. The Viking program, with two JPL orbiters carrying two landers designed by the NASA Langley Center, was a great technical success in summer 1976; all four spacecraft worked. The landers focused on detection of life, but despite some controversy, the consensus of the scientific community was that no life was detected. That had the ironic effect of killing public and scientific interest in Mars exploration for a generation.[10]

Meanwhile, Mariner 10 reached Mercury in 1974 through a Venus flyby—the first spacecraft to fly a "gravity assist" trajectory. Pioneers 10 and 11 made it to Jupiter and Pioneer 11 to Saturn by using Jupiter. The capstone was the Voyager Program. Two spacecraft launched in 1977 reached those two planets between 1979 and 1981; Voyager 2 subsequently flew by Uranus in 1986 and

Figure 5 Carl Sagan stands with a model of the Viking landers that in 1976 made the first successful Mars landings and demonstrated that there were no easily detectable life-forms there. Sagan was one of the key mission scientists, but was also the most important popularizer of spaceflight, astronomy, and extraterrestrial life after 1970. *Source:* NASA/Jet Propulsion Laboratory

Neptune in 1989. They, like the two Pioneers, were going so fast they were on trajectories to leave the solar system. In 2012, Voyager 1 detected the end of the sphere of influence of the solar wind and measured the particles and magnetic fields between the stars. All of these missions were stunning achievements of the United States, which reached every major planet except Pluto (later demoted to dwarf planet) by the end of the Cold War. All these flights produced a wealth of information about the planets, and the icy satellites of the gas giant planets, greatly expanding knowledge about the origins and evolution of the solar system.

Space Astronomy

The idea of a telescope in space is an old one, as it became obvious in the nineteenth century that Earth's turbulent atmosphere limited seeing. Several early space pioneers imagined such a telescope, always as a human-operated observatory in space. Given the state of the technology before 1950, they could scarcely imagine a remotely operated telescope, nor did they anticipate the revolutionary effects on the discipline that would come from opening the entire electromagnetic spectrum, from highly energetic gamma rays to long-wavelength radio frequencies.

Space astronomy, like space physics, began with postwar V-2 flights. A Naval Research Laboratory experiment in October 1946 captured the first images of the Sun's ultraviolet spectrum at wavelengths blocked by the atmosphere. In the 1950s, with smaller, cheaper sounding rockets and recoverable payloads, scientists made tentative explorations of the Sun and the sky in ultraviolet and X-rays. But instruments were small, image resolution was poor, and pointing accuracy was low, so at most one could only gather crude survey data about what was emitting at those wavelengths.[11]

As in other sectors, the space race was transformative, because it suddenly freed up government money to build astronomical spacecraft that never would have been funded otherwise. The United States took the lead, while the Soviets invested less, perhaps due to a lack of resources or political priority. By the 1970s, European and Japanese space science institutions became significant players too.

Soon after its formation, NASA drew up plans for a series of space observatories: geophysical, solar, and astronomical. The solar spacecraft included a section that pointed continuously at the Sun, providing more accurate data about its very hot outer atmosphere and the storms and outbursts that took place on or above its apparent surface. The Orbiting Astronomical Observatories (OAO) were the largest and most difficult. Universities and companies struggled with the demanding requirements for instruments in new wavelength ranges and control systems that could accurately point a telescope at a single sky location for longer periods. Neither task would have been feasible without the massive Cold War investment in parallel military technologies.[12]

The first OAO failed disastrously soon after reaching orbit in 1966. OAO 2 in 1968 carried two telescopic experiments, one from the University of Michigan and the other from the Smithsonian Astrophysical Observatory co-located with Harvard in Cambridge, Massachusetts, since 1955. Harvard astronomer Fred Whipple had taken a nearly defunct Smithsonian program and expanded it into the world's largest astronomical institution in less than two decades by opportunistically taking on every space-related project he could get. By contrast, directors from large, U.S. ground-based observatories still controlled the astronomical discipline and took little interest in space astronomy or government patronage, something historically unavailable to them.[13]

Astronomers had only belatedly accepted the rise of radio astronomy after World War II, which had been fueled by the availability of radar technology and expertise. Now came an expansion into the entire wavelength range that arose from getting above the Earth's atmosphere. By the 1970s, the advantages of understanding astronomical phenomena, notably at the high-energy ultraviolet, X-ray and gamma-ray wavelengths produced by violent processes in the universe, made a transformation of the discipline inevitable, and healthy.

As both space and ground-based projects grew larger and more expensive, they also forced the astronomical discipline to change its political behavior to ensure U.S. government funding. Advocates for a Large Space Telescope worked to mobilize the coalition among astronomers by arguing for its extremely high resolution and ability to reach deeper into the universe once the atmosphere's blurring effects were removed. Disagreements in the scientific community over the value of a billion-dollar enterprise threatened to kill the project in the early 1970s. Congress, in fact, once deleted all funding. Astronomers got the message that they had to unite and build coalitions with NASA centers, aerospace manufacturers, and politicians whose districts might benefit, if they wanted to have it funded. Planetary scientists had learned the same lesson a little earlier. Both the Viking and Voyager programs were less ambitious versions of NASA projects canceled by Congress or the Nixon administration because of big budgets combined with infighting in the scientific community.[14]

Revived as the Space Telescope, with European Space Agency participation to spread the cost, it was approved in 1977 with a somewhat smaller main mirror. Not coincidentally, the 94-inch (2.4 meter) size was available from spy satellite contractors working for the National Reconnaissance Office—and NASA chose the same contractors, Lockheed and Perkin-Elmer, to build the telescope. That size could also be more easily accommodated in the payload bay of the space shuttle NASA was then developing. The agency was committed to trying to make the shuttle the nation's sole launch system, but it also had the ambition to make astronauts into workmen who could repair and upgrade satellites in low Earth orbit. That was an expensive strategy, but ultimately it saved the Hubble Space Telescope (HST), as it was later called, and gave it a much longer, more scientifically productive life.

Solar astronomy also took a path shaped by the human space program. NASA canceled a more advanced solar spacecraft in the mid-1960s in favor of the Apollo Telescope Mount, which would be attached to what became the Skylab space station launched in 1973. It was years late and much more expensive than the robotic version, but astronauts on the three crews on the station did gather pioneering high-resolution images of the Sun over a variety of wavelengths. It set the agenda for solar science for the rest of the 1970s by providing new insights into how the outer regions of the Sun's atmosphere worked. The next large spacecraft was the Solar Maximum Mission (SMM), launched in 1979 to meet the latest peak in solar activity, an objective missed on the previous cycle because of all the delays in Skylab. SMM was designed with modularity so that astronauts could maintain it. That came in handy when it failed in less than a year. In 1984, shuttle astronauts fixed it, and gained experience that would later be valuable for the Hubble, but in this case it would have been cheaper to launch more spacecraft than to spend the hundreds of millions of dollars required for a shuttle launch to fix one satellite.[15]

The 1980s also saw a major expansion of astronomy in the infrared range, below the wavelengths seen by the

Figure 6 The Hubble Space Telescope being deployed from space shuttle *Discovery* on April 25, 1990. The HST turned out to have serious flaws, but astronaut repairs and upgrades soon made it into the most productive and important space science spacecraft ever put into Earth orbit. It has made fundamental contributions to understanding the origin and evolution of the universe. *Source:* NASA

human eye, and typically emitted by cooler objects in the solar system and universe, such as red stars, dust clouds, asteroids, and comets. The Cold War had fueled the development of infrared detectors for missiles and satellites, including versions of the charge coupled device (CCD) on a silicon chip that would later displace film in cameras in civilian use. But it took some time to have those detectors declassified for space science use, so that infrared astronomy came to fruition later than high-energy work. In 1983, NASA launched the Infrared Astronomical Satellite, with major contributions from the Netherlands and Britain. It provided the first all-sky survey of objects visible in that range, and it discovered several asteroids and comets, stars with dusky disks around them, and distant galaxies whose light was shifted into the infrared by expansion of the universe. The success of this spacecraft led to NASA and the ESA pursuing new infrared telescopes to be launched in the 1990s and beyond.[16]

By the 1980s, the contributions of space telescopes, driven by the Cold War space race and bolstered by military technology development, had helped transform astronomy and astrophysics. Major investments in ground-based optical and radio observatories were equally important, and those telescopes also benefited from Cold War expenditures and technology development. As a result of this combined ground and space attack, our understanding of the origins and evolution of the Sun, the solar system, and

By the 1980s, the contributions of space telescopes, driven by the Cold War space race and bolstered by military technology development, had helped transform astronomy and astrophysics.

the universe markedly improved, opening a whole new set of questions to be investigated.

The Life and Earth Sciences

The life sciences in space were dominated from the outset by two unrelated purposes: the impact of spaceflight on living beings and the search for extraterrestrial life. The human space program drove the first and was the prime motivator in early flights of animals as research subjects. As mission durations grew from hours to weeks in the 1960s and 1970s, attention moved from basic questions about whether astronauts and cosmonauts could tolerate spaceflight to a focus on the impact of zero gravity on the human body. As early as the Gemini missions of 1965, it became clear that the prolonged weightlessness caused the body to lose calcium from the skeleton, among other concerns.

Long-duration research into zero gravity's effects, and the impact of exercise and other methods to ameliorate them, became a major area of interest in the two superpowers' space station programs from the 1970s on. After Skylab's three missions of up to eighty-four days, the United States was confined to shuttle flights of no more than eighteen days, although a European-built Spacelab module could be put in the payload bay, allowing

The life sciences in space were dominated from the outset by two unrelated purposes: the impact of spaceflight on living beings and the search for extraterrestrial life.

an increasingly sophisticated series of experimental biology missions that were not only focused on humans. By contrast, the Soviets were flying Salyut and then Mir missions of many months by the mid-1980s, and in the 1990s, cosmonauts flew for up to a year—missions unmatched until the International Space Station was occupied after the year 2000. The result was a trove of medical data on human adaption to zero gravity.

The search for extraterrestrial life was a topic of equally great interest, but it was, for the most part, a minor expense for NASA and other space agencies. The simple reason was that there was nothing to study, since no life was discovered early in the space race. In the United States, the first formal name for this activity was exobiology, but many biologists jibed that anyone who went into that field had become an ex-biologist. That was not fair to experiments that had been conducted since the early 1950s on how the chemistry of the early Earth might favor the formation of life, but indeed a lot of the field was speculative. NASA's first major expenditure was for the biological packages on the Viking landers launched in 1975. These exquisitely miniaturized experiments yielded negative or frustratingly ambiguous results that seemed likely to have been produced by nonbiological surface chemistry on Mars. In hindsight, looking for Earth-like, single-cell life in desiccated soil bathed in solar ultraviolet light and cosmic radiation seems optimistic at best, but it is a clear indication of where the field was in the 1970s. A new turn came in the 1990s, as discoveries of "extremophiles"—exotic forms of Earth life living in seemingly uninhabitable environments such as in superheated, acidic water around undersea volcanic vents—opened the field to new possibilities. Sensing the need for a rebranding, NASA began calling the field "astrobiology."[17]

The earth sciences evolved primarily from practical concerns, like predicting the weather and better managing use of the land and sea. NASA's atmospheric and oceanographic research series, Nimbus, followed the first weather satellites in the 1960s. The Landsat series began in 1971 and demonstrated the value of multi-spectral imaging of land surfaces. Soviet, European, and Japanese spacecraft followed. As robotic spacecraft proved themselves able gatherers of science data, they fueled the rise of "Earth systems science" in the 1980s. Planetary exploration contributed as well. Remote-sensing instruments benefited from technology development in that program, and Venus and Mars atmospheric data helped stimulate a comparative planetology, shedding light on the Earth's evolution and global processes.[18]

Politically, increasing worry about the impact of pollutants on the ozone layer, which protects the Earth from solar ultraviolet radiation, and about the effects of greenhouse gases on global climate, fueled rising budgets. NASA began formulating "A Mission to Planet Earth" in 1987, at first with massive, possibly human-tended, platforms. But smaller, more budget-conscious satellite projects blossomed in the 1990s and beyond, supplemented by spacecraft from other nations. Like the other major space science fields, the growth of multiple, intertwined programs and institutions in earth science fostered a vibrant global community, with both formal, international cooperative programs and informal, transnational exchanges among scientists, institutes, corporations, and agencies.

Planetary Science and Astronomy in the Late and Post–Cold War Years

Although the Voyager flybys of the outer planets made it appear that exploration was flourishing, the 1980s actually were troubled times for NASA's planetary program. Weak budgets in the late 1970s were followed by the Reagan administration's 1981 attempt to cancel the entire program, spin off JPL, and give the money to the space shuttle. Congressmen representing threatened centers or contractors helped fend off those scenarios. As it was, the national policy of forcing all payloads onto the shuttle had already caused major delays and budget increases for the few missions in development. After the *Challenger* disintegrated in 1986, all launches were delayed three years, so NASA sent no new planetary spacecraft skyward between 1978 and 1989. In a blow to U.S. prestige, the agency could not afford to intercept Comet Halley during its high-profile 1985–1986 return, while the Soviets flew by with a variant of its large Venus vehicle and the European Space Agency launched its first interplanetary probe, Giotto, taking the first close-up images of the icy nucleus of a comet.[19]

The Hubble Space Telescope, which was already late and over budget due to its complexity, was delayed four more years by *Challenger*. That was a blessing in disguise, since it could have failed completely if key systems had not been upgraded during the wait. When Hubble was orbited in 1990, it became a national embarrassment. The main mirror turned out to have been ground precisely to the wrong figure, making images a little blurry. NASA had eliminated a second optical testing method in the early 1980s to save money. It was still a usable scientific instrument, but it damaged the agency's credibility, at least until shuttle astronauts carried out a remarkable repair mission at the end of 1993.[20]

On the other hand, astronomy also was the beneficiary as Republican administrations increased NASA's budget in the late 1980s. The agency had successfully packaged Hubble and three other missions as the Great Observatories. These became the Compton Gamma Ray Observatory (launched 1991), the Chandra X-Ray Observatory (launched 1999), and the Spitzer Space Telescope (formerly the Space Infrared Telescope Facility) (launched 2003). Smaller missions from multiple nations also flourished in the 1980s and 1990s. On the Soviet side, they orbited two mid-sized, high-energy telescopes during the 1980s and flew cosmonaut-operated instruments attached to their Mir space station.

Unfortunately, Russian space science capability went into a steep decline with the economic crisis and breakup of the Soviet Union. No more astronomy missions left Earth until the 2000s, and the planetary program fell apart as underfunding robbed the Lavochkin design bureau of capability. In 1988, the Soviets launched two Phobos spacecraft for Mars imaging and close encounters with the Martian moon of the same name. One failed in transit and the other died in orbit shortly before it reached the little moon. A similar Phobos mission was attempted in 1996, with NASA and ESA cooperation, and another in 2011, with Chinese participation, but both were lost during launch due to poor quality control in Russia's space industry. The country's leaders focused instead on keeping their human spaceflight program alive, propped up by U.S. money after the two nations merged their space station programs in 1993–1994.[21]

If the Cold War's end was devastating for post-Soviet space science, its effects on the United States were more subtle. With the space race over, the agency's budget went into a gradual decline during the 1990s, exacerbated by the Hubble embarrassment and a reputation for having become slow and bureaucratic. Yet the degree to which American space science continued to prosper indicates that it had created its own institutional and political momentum. Space centers, university and nonprofit institutes, corporate contractors, and agency bureaucracies depended on continued government funding, and politicians were happy to have those well-paid, high-technology jobs in their districts. Investments in space science also sustained national technological prowess, with defense implications, and successful missions signaled that prowess and brought international prestige.

Nonetheless, the combination of political dissatisfaction with NASA's bureaucracy and budget pressures forced the agency to reform its space science programs in the 1990s. In 1992, the first Bush administration replaced NASA Administrator Richard Truly, an ex-astronaut, with Daniel Goldin, an engineering executive from a major defense contractor. He lasted through both terms of Bill Clinton's presidency because of his shake-up of the agency. Soon labeled "better, faster, cheaper," Goldin's program aimed to impose leaner organizations, less paperwork, fewer review boards, and more risk-taking, based on experiences in President Reagan's short-lived, 1980s initiative for space-based missile defense.[22]

The impact on NASA's planetary program was especially noteworthy. Reform initiatives had begun before

Goldin due to the unhappy experiences of the 1980s. Two large spacecraft projects, the Magellan radar mission to reveal the surface of cloud-shrouded Venus, and the Galileo orbiter and atmospheric probe to be sent to Jupiter, had consumed much of the planetary science budget for years. NASA's shuttle policy and the *Challenger* accident added further expense and delays, although both spacecraft succeeded after suffering serious in-flight failures. Meanwhile, an attempt to create a lower-cost spacecraft ran over budget and produced only one vehicle, Mars Observer, which blew up just before it reached the planet in 1993.

A second cost-reduction attempt was much more successful. The Discovery Program, which Congress approved that same year, opened up planetary exploration to cheaper, more innovative missions chosen after competitions between science and engineering teams. Johns Hopkins's APL became JPL's principal rival and built the first Discovery spacecraft launched, Near Earth Asteroid Rendezvous. It flew by a small asteroid, orbited, and finally landed on the asteroid Eros in 2000–2001. JPL's Mars Pathfinder, having a much shorter trajectory, pulled off an innovative, airbag-cushioned landing in 1997, and deployed a mini-rover. It was the first successful vehicle to reach the Red Planet in two decades.[23]

The next two JPL Mars missions, however, failed embarrassingly in 1999, showing that doing risky things on the cheap had its limits. Goldin's "better, faster, cheaper" program suffered a political setback, made him risk-averse, and accelerated the end of his tenure at NASA. Those two Mars missions were not part of Discovery, but that program went into crisis too in the early 2000s, thanks to the overly optimistic assumptions about the cost and schedule required to do more ambitious things, like hit a comet, or put an orbiter around the planet Mercury. Nonetheless, Discovery produced a string of flights, notably to comets and asteroids, and proved the value of competition. NASA extended the idea to medium-sized missions and in late 2001 gave the Southwest Research Institute and APL the job of flying by Pluto, which the New Horizons probe reached in 2014. Meanwhile, NASA's Mars program gathered great momentum after the 1999 failures, with a series of orbiters and three rovers exploring the surface, two of which, Opportunity and Curiosity, are still active as of this writing.[24]

Despite America's unparalleled investment and success in planetary exploration in the post–Cold War era, it is noteworthy how much Europe and then Asia emerged as serious players, especially after 2000. NASA's Cassini Saturn orbiter, launched in 1997, carried a European probe that landed on Saturn's moon Titan in 2005. ESA Moon, Mars, and Venus orbiters all succeeded between 2003 and 2006, and the agency collaborated with Russia to put a new spacecraft around Mars in 2016. The agency's Rosetta

Despite America's
unparalleled investment
and success in planetary
exploration in the
post–Cold War era,
it is noteworthy how
much Europe and
then Asia emerged
as serious players,
especially after 2000.

spacecraft reached a comet in 2014, and it dropped a small lander on its surface. Japan, which began with a small particle-and-fields mission to fly by Halley's Comet in 1986, succeeded with Moon, Venus, asteroid, and comet missions in the 1990s and 2000s. China's lunar exploration program began with Chang'e 1 in 2007, and in 2015 Chang'e 3 landed on the Moon and deployed a small rover. India succeeded in orbiting the Moon in 2008 and Mars in 2012.[25]

A similar story unfolded in space astronomy after the Cold War. The United States, with a civilian space budget larger than all others' space budgets, funded an array of important missions, while the Europeans and Japanese began to catch up. Astronaut repair, maintenance, and upgrade flights turned the NASA-ESA Hubble from a late-night TV joke into a popular vehicle for gathering fundamental data about the universe. Especially noteworthy was the establishment, in conjunction with ground-based observatories, of a fairly precise measurement of how fast the universe is expanding, which led to a determination of the time since the Big Bang: 13.7 billion years. The other Great Observatories produced a rich haul of data on black holes, exploding stars, dust clouds, and the very early evolution of galaxies. NASA small and medium-sized telescopes added specialized contributions, such as mapping the irregularities in the faint background radiation the Big

Bang produced, showing the seeds of the earliest development of galaxies and stars. Following a technological breakthrough in ground-based instruments that conclusively demonstrated for the first time that planets indeed orbited around other stars, NASA's Kepler telescope found thousands more planets and is continuing to do so today.

ESA contributed to all fields as well, launching pioneering missions like Hipparcos and Planck to precisely measure the positions and motions of millions of stars in our galaxy, yielding precise distance measurements and a much better understanding of the Sun's neighborhood. Beginning in 1979, Japan began orbiting astronomical satellites too, notably specializing in infrared and X-ray astronomy. Economically less developed China and India have done less, as they have prioritized practical civil or military applications, and missions that garner international prestige, like flights to the Moon and planets.

Finally, large investments by NASA, ESA, and Japan supported a growing network of satellites to continuously monitor the Sun and its outbursts. Some were parked in an orbit a million miles from Earth, at a special balance point between terrestrial and solar gravity, allowing twenty-four-hour observation of solar flares and ejections of charged particles that would impact the Earth's magnetosphere. Such "space weather" events create beautiful auroras (northern and southern lights), but also threaten both the growing infrastructure of satellites in Earth orbit and, through currents induced by magnetic storms, also ground infrastructure like electrical power networks. Space-based solar astronomy and space physics, which shared a common origin in V-2 flights after World War II, thus became increasingly integrated into a multidisciplinary field that NASA branded "heliophysics." Data from particles and fields experiments on Earth-orbiting and interplanetary spacecraft of multiple nations are being combined with ground and space solar observations, creating for the first time the outlines of a comprehensive vision of the Sun's influence on the entire solar system.

Conclusions

What has humankind learned from nearly seventy-five years of space science and exploration? As the last example indicates, an enormous amount. Powered by an unforeseen robotic spacecraft revolution, but sometimes also by direct human exploration, and paired with ground-based observatories and laboratories of increasing sophistication, space science has produced a growing understanding of the evolution of our planet, solar system, galaxy, and universe. We also have begun to gather the data on the possibilities and risks of sending humans deeper into space, and have set the foundation for the discovery of extraterrestrial life, perhaps in the nearer future. Nothing coming out of the sudden achievement of spaceflight is more profound than that. But space technology has also served another purpose: the transformation of life on Earth, for good and ill, through satellite networks that have become essential to our daily existence.

4

A GLOBAL SPACE INFRASTRUCTURE

Out of the Cold War space race also came infrastructures, both military and civilian, that quickly became critical to life on Earth. Networks of satellites and ground stations produced weather and navigation data, global communications and television, intelligence imagery and information, military command and control, early warning of rocket launches, and other applications that governments and many individuals, especially in the developed world, could no longer imagine living without. These satellite systems were another product of the robotic spacecraft revolution that allowed so much to be done in space remotely.

One important effect of this military and civilian infrastructure was to increase strategic stability between the nuclear-armed superpowers: knowing what the other was doing made war less likely, including war in space.

Moreover, both sides became dependent on their respective satellite networks, further motivating them to accept the de facto regime that space could be militarized, but not weaponized. While Soviet space weapons tests and President Reagan's SDI threatened to destabilize that regime in the 1980s, the end of the Cold War produced an era of U.S. global hegemony that favored the continued avoidance of space weapons—although that is by no means guaranteed for the future.

The rapid emergence of space infrastructures had another effect: by the 1980s, humankind had effectively annexed near-Earth space to serve the planet. Most satellites came to be concentrated in three zones: (1) geostationary Earth orbit (GEO) at about 22,300 miles, dominated by global communications and observation satellites; (2) medium Earth orbit (MEO) at about 11,000–12,000 miles, occupied primarily by navigation spacecraft; and (3) low Earth orbit (LEO) at around 1,200 to 100 miles, with payloads in orbits at all inclinations doing almost everything. In practice, most satellites in that zone orbit between 200 and 600 miles. Above 600 miles is the strongest part of the inner Van Allen radiation belt, making that region undesirable for long-lived vehicles. Below roughly 200 miles, drag from very tenuous atoms of Earth's outer atmosphere bring satellites down too quickly, although high-resolution reconnaissance satellites may dip much lower to get the most revealing views.

As was true in other space sectors, the post-Cold War, somewhat temporary decline in Russian capability was more than compensated by the continued rise of Western Europe, Japan, Canada, China, and India. That meant that civilian and military space infrastructure continued to globalize even as space networks fed the globalization of the world.[1] Moreover, the ideology of free markets that got renewed prominence in the West in the 1980s accelerated the commercialization of civilian space systems. Multinational corporations, rather than governments or government-sanctioned monopolies, came to control a large fraction of satellite communications, and orbital launch and Earth imaging services partially migrated into the commercial sector.

In this chapter, I group space infrastructures into three broad categories. The first two, observation and communications, became the dominant Earth-focused applications of the early Space Age. But a noteworthy phenomenon of the 1980s and later was the growing global dependence on location and timing services from government navigation satellites originating in the military. Navigation thus became a critical third category. As the U.S. Air Force–run Global Positioning System (GPS) came to be integrated into automobiles and smartphones, it became an important sphere of commercial activity and daily life, deepening our dependence on space infrastructure. Consequently, Russia, China, Western Europe, and others began to develop competing systems. For the public, these systems, like their counterparts in the first two categories, are effectively invisible: one only notices space infrastructure on the rare occasions when it does not work.

Observing Earth

As in all aspects of rocket and space technology, the line between military and civilian Earth observation was blurry from the outset. Reconnaissance satellites, which had been so critical to the origins of the space race, needed high-resolution imaging. But lower-resolution cameras, sometimes derived from failed intelligence projects, were suited for taking pictures of weather systems or lunar and planetary surfaces. Weather information was equally valuable to military and civilian clients. Military and intelligence requirements often required a considerable investment in secret sensor development—for example, in the infrared region for detecting the heat signatures of enemy rocket launches. But as weather and Earth science satellites exploited more and more wavelength bands, they needed advanced instrument development too, often derived from declassified military technologies.

In the United States, the National Reconnaissance Office flew CORONA spy satellites, which returned photographic film in reentry capsules, until 1972, supplemented

As in all aspects of rocket and space technology, the line between military and civilian Earth observation was blurry from the outset.

Figure 7 The first space reconnaissance photo ever taken shows an airfield in the Soviet Arctic on August 18, 1960. The first successful U.S. CORONA spacecraft to carry a camera, flying under the cover name "Discoverer 14," made the image. *Source:* National Reconnaissance Office

by similar, high-resolution GAMBIT satellites to take close-up images of targets. Electronic methods of returning images could not produce the resolution required to be militarily useful before the late 1970s. Meanwhile, NASA launched the first experimental weather satellite, TIROS 1 (Television Infrared Observation Satellite), in April 1960, a simple spinning vehicle with TV cameras that had evolved from a losing reconnaissance proposal by Radio Corporation of America (RCA). Like CORONA, what was to be a temporary stopgap or experiment became the basis for an operational system of satellites orbiting over the poles so that all of the globe would be visible.[2]

At first, meteorologists needed to be convinced that images that showed mostly clouds would be useful, but most quickly concluded that it provided valuable information on weather fronts and storms. Detecting and tracking hurricanes in the open ocean greatly increased warning time. When TIROS images did not prove as useful as hoped in predicting weather over the Soviet bloc, the U.S. Defense Department built a parallel system based on the same technology. Its first and most important job was producing cloud cover predictions that would allow spy satellite operators to know when to attempt photography, so as to reduce the number of pictures of overcast targets.

The Soviets deployed their first Zenit reconnaissance vehicles in 1962, using a version of the Vostok spacecraft, as Sergei Korolev had planned from the outset. The spherical reentry module, instead of carrying the cosmonaut and his ejection seat, held the cameras, which were recovered along with the film. But the early Soviet emphasis on producing propaganda spectaculars with humans or lunar and planetary probes meant that the USSR lagged behind the United States in every practical application of space technology. The first experimental Meteor weather satellites using television cameras were not launched until 1964, and all were hidden under the generic Kosmos label until the system was declared operational in 1969.[3]

While the United States led the way in weather systems, it struggled with how to organize responsibility for civilian programs. NASA as an experimental and developmental agency was not the best organization to run an operational system, but it was the agency on the cutting edge of civil spacecraft and sensor development. It pursued its own Nimbus satellite project even as it transferred the simpler systems evolved from TIROS to the National Weather Service, which eventually became part of the National Oceanic and Atmospheric Administration (NOAA). Nimbus did lead to the development of instruments in infrared and microwave regions that could greatly expand what weather satellites could do. It was only when satellites could make comprehensive measurements of water vapor, cloud layers, temperatures, and other quantities, in the 1980s and later, that they began to produce data that could be fed into increasingly powerful computer models for weather prediction.[4]

NASA also pioneered the stationing of weather sensors in GEO. In a twenty-four-hour circular orbit at 22,300 miles over the equator, a satellite effectively hovers over one place. Agency experimental spacecraft in the late 1960s led to the creation of a network of geostationary weather monitors that provide a global overview of weather systems and developing storms that supplements the close-up view from polar-orbiting satellites in LEO. NASA launched spacecraft in 1974 and 1975, leading to the NOAA-operated Geostationary Operational Environmental Satellites (GOES). Through international meteorological and atmospheric organizations, the United States, Japan, and the European Space Agency created a global system in the 1970s, with the ESA satellite responsible for observing Europe and Africa, and the Japanese one responsible for East Asia and the Western Pacific. The Soviet Union promised to put up an Indian Ocean satellite, but the responsible design bureau struggled with technological challenges and indifference in the military-run space program. It was only in 1994 that a post-Soviet Russia put up its geostationary weather spacecraft, and it was plagued with technical troubles. Thanks to the country's economic crisis, no replacement was launched until 2011.

Driven by the Cold War, military and intelligence observation systems also came to encompass a range of wavelengths and techniques far beyond optical photography. The first successful U.S. intelligence satellite was actually a navy signals payload orbited covertly on an NRL astronomy spacecraft in May 1960. Its role was to record Soviet air defense radars, enabling reconstruction of their locations, frequencies, and strength. Signals and communications intelligence capabilities from orbit vastly expanded over time, but very little information about that has been declassified. Experiments with radar imaging (using a radio signal reflected off the ground to form an image) began in the 1960s, but due to the difficulty of obtaining good pictures, the first U.S. Lacrosse operational radar-imaging satellite was not launched until 1988. The advantage of radar is that it can image at night and through clouds. Unique data may also be derived from the target's reflection characteristics.

The United States also began experimenting with infrared sensor missile early-warning spacecraft in the mid-1960s. That led to a geostationary system, first completed in 1973, with three satellites, one watching the Soviet Union, and one each over the Atlantic and Pacific to detect submarine-launched ballistic missiles. This considerable increase in warning time over ground-based radars could enable better nuclear warfighting, but it also reduced the fear of surprise attack. The Soviets followed eventually with their early-warning spacecraft in the 1980s. These mostly used a highly elliptical orbit, in which the spacecraft ascended high over the Northern Hemisphere during the slow-moving part of its orbit, then whipped around the Southern Hemisphere while it descended to a few hundred miles at the lowest point. But in this as in all systems, the Soviets struggled with the American technological lead, notably in the size, weight, and reliability of electronics and computers. Soviet vehicles often had abbreviated lifetimes, and in 1983 one issued a false warning of American missile attack when sunlight reflected off clouds. An alert technician prevented Armageddon. After the collapse of the USSR in 1991, the system deteriorated for a couple of decades, and the Russians had always depended on it less, but on balance it probably also contributed to strategic stability.[5]

Photo reconnaissance systems were the most crucial, however, as they made the first U.S.-Soviet nuclear arms control agreements feasible in the 1970s. Missile silos, bomber aircraft and nuclear submarines (when in port) could be counted, providing the basis for treaty limits. As a result of declassification, much more is known about optical U.S. systems, at least until the 1980s. The NRO and the Air Force replaced the CORONA series with Hexagon satellites, beginning in 1971. Often called the KH-9, after its main camera, and nicknamed Big Bird because it was so huge, each HEXAGON carried four larger reentry vehicles to return film of broad areas of the Soviet bloc, China, and other adversaries. High-resolution satellites like GAMBIT and its successors would then investigate particular targets.[6]

Film return limited the utility of photo reconnaissance spacecraft primarily to strategic intelligence. No images could be returned quickly enough to aid U.S. forces in a crisis or war like Vietnam—unlike defense weather satellites. Beginning in 1976, the United States orbited the first digital imaging spacecraft, using early silicon chip CCDs, now the basis for every commercially available camera. The KENNEN (later CRYSTAL) vehicles (often publicly identified as the KH-11) resembled the Hubble Space Telescope, which had been derived from them. Transmitting images to the ground solved the problem of timeliness and abbreviated spacecraft lifetimes (once film return spacecraft deployed their last reentry vehicle, they were finished). Yet the transition was not an instantaneous one for the NRO, as HEXAGONs were orbited until 1984. Little is known about the KENNEN series in public, but the end of film return spacecraft certainly signaled that electronic imaging was now adequate for both broad area search and close look missions. Supplemented by radar imagers and an unknown variety of signals intelligence payloads and satellites, the United States had gained a capacity for rapid global surveillance that bolstered its role as a late and post–Cold War hegemonic superpower. Still, as the decades since have proved, quasi-omniscience at the strategic level is no guarantee of military or political success when the problem is intractable or the enemies are guerilla and terrorist groups.[7]

Even as the Soviets began using electronic imaging spacecraft in 1982, they struggled to make them work and remain competitive. To keep America and its allies under observation, they and their Russian successors launched many short lifetime, film return spacecraft with increasingly advanced camera systems. One unique Soviet vehicle was a Radar Ocean Reconnaissance Satellite (RORSAT) using a nuclear reactor for electrical power, in order to track the global fleets of the U.S. Navy and its allies.[8] With the post-Soviet economic crisis, Russia lost a lot a global reconnaissance capability for two decades, only gradually reconstructing it after 2010.

Despite American predominance in Earth observation, the 1980s began a rapid shift toward a multipolar world in which intelligence agencies no longer had a monopoly on high-resolution images. U.S. intelligence agencies had enforced a national policy keeping all civilian spacecraft, including weather satellites and the Earth resources Landsats, to a resolution no better than 30 meters (about 100 feet) per pixel. In 1986, the French space agency, in cooperation with Sweden, orbited SPOT 1, with 10- to 20-meter resolution photos available for a price. It was the beginning of a civilian revolution in imaging, which by the 2000s led to companies like Digital Globe selling increasingly sharp images on demand. Even the media and human rights groups could begin to buy pictures of interesting targets, and companies like Google could make satellite imagery central to their map software. Intelligence agencies began to buy large quantities of commercial images to supplement their own specialized satellites. And multiple nations began launching their own military reconnaissance spacecraft, notably France and Germany, Israel, and China. LEO and GEO orbits became saturated with observation spacecraft of many different types. The availability of so many satellites created a new era of global transparency, reducing the chance of major wars, without by any means removing reasons for conflicts around the world.[9]

Satellite Communications

Like reconnaissance satellites, ideas for using spacecraft for global communications go back to the end of World War II and the rocketry revolution unleashed by the V-2. In October 1945, science fiction author and space advocate Arthur C. Clarke, then a junior radar officer in the British Royal Air Force, proposed that three geostationary platforms could provide global communications. Given the technology of the time, he imagined them as human-tended stations, because, of course, someone would be needed to change the vacuum tubes. Clarke's concept was prescient, but it was exotic and ignored for some time. When the space race opened in the late 1950s, the U.S. military services, NASA, and private corporations experimented with several different concepts for robotic systems, including giant balloons or a belt of copper needles to reflect signals, low Earth orbit "store and dump" communications vehicles, and geostationary concepts. Telecommunications giant AT&T planned a LEO constellation based on its Telstar satellite launched in July 1962. The broadcasts between Europe and North America that Telstar relayed made a profound impression on the world, even though each lasted only about fifteen minutes as the spacecraft soon passed out of range. It heralded a new world of instant global communication.

Even before Telstar was launched, however, AT&T's concept for a privately operated LEO system was already losing out. The Kennedy administration, more interested in beating the Soviet Union to a global system, wanted to strengthen American influence over the West and the developing world. Parallel to the Apollo decision, the administration began a study of space communications in 1961 that led to a law the president signed at the end of August 1962. It established a public-private monopoly, the Communications Satellite Corporation (COMSAT). AT&T became just one of the stockholders. NASA had also funded competing experiments, notably the Hughes Aircraft–built Syncom (synchronous communications) spacecraft for GEO. Syncom 2 and 3 in 1963 and 1964 proved the feasibility of the geostationary robotic satellite concept. COMSAT meanwhile had initiated a set of global negotiations, as set out in the act, leading to the creation of the International Telecommunications Satellite Organization (Intelsat) in 1964, in cooperation with the mostly government-owned telephone companies of Western and developing nations. Dominated by COMSAT, it adopted a version of Hughes's Syncom. The first of its spacecraft, Intelsat I or Early Bird, was positioned over the Atlantic in April 1965. Higher-capacity second-generation satellites followed in 1966–1967, third-generation in 1968–1970, and fourth-generation beginning in 1971. Due to satellite failures, it was really at the beginning of the 1970s that permanent global coverage emerged.[10]

The Soviets naturally rejected this American-dominated system and kept its allies out of it. They launched their first experimental, military Molniya (Lightning) communications satellites, or comsats, in 1964 and 1965, pioneering the highly elliptical orbits later adopted by their missile early-warning vehicles. Geostationary satellites were too close to the horizon in the USSR's Arctic regions, and equatorial orbits were harder to reach from high-latitude Soviet launch sites and would require new booster development. So highly elliptical polar orbits in which the spacecraft lingered over the Northern Hemisphere were a good solution, but a solution that required

Figure 8 Intelsat 1, nicknamed "Early Bird," became the world's first commercial, geostationary communications satellite when it was parked over the Atlantic in April 1965. In this publicity photo, a model spacecraft hangs over an array of telephones, as long-distance calls were Early Bird's primary function. Over time, undersea cables took over telephony and commercial satellites dominated long-distance television transmissions. *Source:* Smithsonian National Air and Space Museum (NASM 7B14138)

multiple Molniya satellites in different orbital planes so that at least one was overhead at any given time. Given the military-controlled unitary space program of the USSR, military users came first, but the spacecraft were also employed for civil television broadcasts across the vast expanse of Soviet territory. In the later 1970s, the Soviets also began to deploy geostationary satellites as a supplement.

The U.S. Defense Department, after briefly deciding to rely on the civil system, began launching its own comsats in 1966 and established a full geostationary system beginning in 1971. A key consideration was maintaining a sufficient level of security for critical command and control messages, particularly while fighting a global nuclear war. "Urgent action messages"—presidential commands to release nuclear weapons—needed to reach ballistic missile submarines at sea and bases overseas, reflecting the global scale of American military deployments. Satellite communications thus could help bring about the end of the world, but as with every other aspect of the Janus-faced logic of nuclear weapons, it also strengthened the credibility of the U.S. deterrent and lessened the likelihood of war. As defense communications capability grew, it inevitably also led to a growing reliance on it for all levels of messaging, not just the most critical. The U.S. Defense Department and the armed services began launching a variety of satellites that took over a part of the burden formerly borne by ground systems, long-range radio, and undersea cables. That did not prove to be enough, so rental of commercial circuits also grew, providing a major source of business for private satellite companies. By the end of the Cold War and beyond, U.S. global hegemony came to be dependent on space infrastructure in communications, but also observation and navigation, creating a vulnerability that has encouraged some USAF-affiliated strategists to argue that America should be deploying orbital weapons to dominate near-Earth space—likely setting off a new arms race.[11]

Cold-War–driven government initiation and ownership of much of the communications satellite infrastructure, however, should not obscure the fact that it became the first—and for at least two decades the only—sector where a profit could be made on space activities not funded by governments. Money earned from carrying telephone calls and TV broadcasts would eventually allow telecommunications companies and new enterprises to break out of the monopolistic COMSAT/Intelsat system. Domestic communications became the first independent area; in 1971 Canada created the second national system after the USSR. The next year, the U.S. Federal Communications Commission declared an "open skies policy" that allowed companies to launch spacecraft for domestic service, reflecting primarily the lower cost of transmitting television coast-to-coast that way. While satellite telephone calls were never competitive domestically and soon fell internationally because better undersea copper and fiber-optic cables became available, global television remained the foundation of the communications satellite business. As satellites became larger and more powerful, direct broadcast to smaller dishes became feasible in the 1980s—to ships at sea and eventually to individual homes.

The expansion of the business, reinforced by the revival of free market ideology in the West in the 1980s and 1990s, led to the breakup around 2000 of COMSAT and Intelsat (and also Inmarsat, a parallel international organization for maritime and mobile users). They became commercial companies with a lot of competition. The 1990s also saw a boom in new low-orbit proposals involving many satellites, notably Iridium, founded by the electronics firm Motorola. Iridium created a global satellite telephone system, justifying it in part by proposing service to developing countries that geostationary systems had failed to deliver. But Iridium was a spectacular market failure, as ground-based cellular telephone systems killed demand for expensive and heavy satellite phones. The company went bankrupt only months after launching its first satellites in late 1998. Tellingly, it was revived in 2001 due to the support of the U.S. Defense Department, which found telephony to remote areas like Afghanistan important. The original owners lost their investment, but Iridium continues to this day as a private corporation, and is even launching a new satellite constellation, thanks to military and some media business.[12]

While GEO systems remain dominant, a new speculative wave of LEO systems has emerged after 2010, based on ideas for spreading global Internet access via spacecraft, among other applications. It is too soon to tell how that will play out, but in a world saturated with communications systems via fiber-optic cables and satellites, and with the expectation that computer network access will soon be available anywhere on or above Earth, space infrastructure will remain critical to global functioning. The political, economic, cultural, and military forces that drive globalization are much bigger than satellite communications, but there is no doubt that the technology has been a major conduit for globalizing business, culture, entertainment, military power, and every other sector.

Space-Based Navigation

Unlike satellite observations and communications, navigation systems were scarcely anticipated before the space race. Immediately after the launch of Sputnik in October 1957, two engineers at the Johns Hopkins APL noticed that tracking the Doppler shift in Sputnik's radio transmission as it moved toward or away from the observer, which was used to determine its orbit, could be turned around to fix a position on Earth, if the orbit was well known. That led to the navy's Transit system, with satellites built by APL, the primary purpose of which was to provide locations to ballistic missile submarines. Its first successful launch was in 1960, and the system became operational in 1964, based on four to six satellites orbiting at 1,075 miles. Transit allowed the inertial guidance systems in submarines to give a launch position precise enough to hit targets in a nuclear war. The navy system was sometimes used by the other services for positioning, and civilian users even adopted it for surveying and other applications.[13]

Transit, however, could only provide a latitude and longitude position on the surface of the Earth, and it could take up to half an hour to compute a position. The navy began experiments to orbit atomic clocks to provide precise time as an alternate method of position determination. The USAF experimented with its own satellites to provide aircraft positions and altitudes. The army and NASA had flown a series of geodetic spacecraft to take measurements of the shape of the Earth and its gravitational field, data critical to improving the accuracy of global maps and of long-range nuclear missiles. In late 1973, the U.S. Defense Department forced a merger between the competing service programs, taking the best technology from each. Called the NAVSTAR Global Positioning System, and now universally known by the acronym GPS, it would be operated by the air force. Beginning in 1978, that service put GPS satellites in approximately twelve-hour circular orbits at 11,000 miles altitude (the zone that came to be called MEO). The system started limited operations in the early 1980s, and reached initial operating capability in 1993, when there were twenty-four working spacecraft in six different orbital planes. It enabled nearly instantaneous, three-dimensional positions and time over the whole globe, with higher-accuracy signals reserved for the U.S. armed forces. It is impossible to imagine this massive investment without the nuclear arms race and the global deployment of U.S. forces, but its value proved so great, also to civilian users, that it in effect became a government-operated national utility, maintained by the United States because it had become too important to life on Earth—paralleling the case of weather satellites.[14]

The Soviets imitated Transit and followed that with the first launch of the GPS-like GLONASS system in 1982. Its satellites were in higher inclinations to provide more accuracy to polar regions. It became operational in 1995, but the economic crisis in post–Soviet Russia resulted in deterioration in capability as failing satellites were not replaced. In the 2000s, however, President Vladimir Putin ordered that the full system be restored and upgraded to be competitive with GPS. GLONASS became the second global system widely integrated into civilian use, such that many receivers like mobile phones use both simultaneously to increase accuracy of position.

Political reasons have impelled the creation of further systems, since the American and Russian ones are military-operated and could potentially be degraded, jammed, or turned off for civilian users in case of a crisis. The European Union began discussing its global navigation satellite system, Galileo, in the late 1990s. Even after President Clinton opened the more precise military GPS signal to all users in 2000, Galileo continued largely because of discomfort in some European capitals with dependence on the U.S. military. The first Galileo test satellite was launched in 2005, but the system is still under construction as of this writing, as are Chinese, Indian, and Japanese ones. The explosion in use by civilian devices in vehicles and cellular phones have made navigation satellites so integral to daily life, especially in the developed world, that it is easy to predict that this expansion of location-based services will continue and lead to innovative new applications.

Space Infrastructure and Globalization

The construction of a spectrum of satellite infrastructures has clearly had multiple and sometimes contradictory effects. It has strengthened the military power of the United States and the Soviet Union, and more recently Western Europe, China, and a resurgent Russia, and it has bolstered the ability to fight a global nuclear war—while simultaneously making such a catastrophe less likely through global transparency. War in space also became possible through the development of anti-satellite weapons, but the very dependence of major powers on infrastructural systems has produced its own deterrent effect: the taboo against the weaponization of space is still intact, if increasingly fragile, as ground-based missile and electronic threats against satellites increase. National power, particularly for a few of the wealthiest, more technologically advanced countries, may have been strengthened, but nonmilitary weather, Earth science, and communications satellites have also promoted globalization. Global television and images of Earth from space, notably from geostationary satellites and Apollo, have contributed to the tentative emergence of a planetary identity.

Commercializing space infrastructure has also fostered transnational businesses, and the transnational circulation of people and knowledge, altering the nationally focused and government-dependent space industries that sprang out of the Cold War. The Satellite Industry Association reported in 2012 that of 994 active Earth satellites, 38 percent were commercial communications satellites and another 20 percent were government and military ones. Global space industry revenue in 2011 was US$298.8 billion, of which $177.8 billion (61%) came from the communications satellite industry (the rest included launch

Global television and images of Earth from space, notably from geostationary satellites and Apollo, have contributed to the tentative emergence of a planetary identity.

services, human spaceflight, non-communications spacecraft, etc.). Much of that 61 percent consisted of ground infrastructure and the sale of TV transmissions.[15] When one considers that the total NASA budget in those years was around $18 billion, it puts in perspective the public and media focus on human flights, plus occasional images from space science missions, as all that matters in space.

One of the by-products of the intensive use of the LEO and GEO zones is the growing danger of space junk. For every working satellite there are thousands of pieces of debris, plus dead rocket stages and spacecraft. A catastrophic cascade of collisions, sometimes called the Kessler syndrome because it was NASA scientist Donald J. Kessler who first outlined it in 1978, could make certain orbits unusable. The upper part of the LEO zone is especially vulnerable, as debris does not come down quickly and satellites orbit at all inclinations to the equator, setting up potential impacts between objects moving thousands of miles per hour. Attacks on satellites by surface-based missiles, which most easily reach LEO objects, might trigger or accelerate the process—a Chinese anti-satellite test in 2007 against one of their defunct spacecraft produced a cloud of thousands of fragments,[16] as did an accidental collision between an active Iridium spacecraft and a defunct Soviet satellite in 2009. Losing LEO would be a major shock to a planet that has become dependent on space services. It could have far-reaching effects on both military effectiveness and daily life.

Conclusions

One of spaceflight's most profound impacts has been the creation of an Earth-orbiting infrastructure. The daily lives of billions of people are now shaped by satellite communications, navigation, and weather information. Global politics and military establishments have become dependent on those systems too, plus intelligence and early warning from space. Infrastructural systems easily outlasted the Cold War space race that spawned them because their utility justified increasing government or business investment. Indeed, satellite infrastructures have shaped the very cultures we live in through the global dissemination of information and entertainment. In the process, the once exotic topic of space travel has become normalized and embedded in popular culture and everyday life.

5

ASTROCULTURE: SPACEFLIGHT AND THE IMAGINATION

In order to make spaceflight real, one first had to imagine it. The rise of space science fiction in the nineteenth century, followed by nonfiction space advocacy in the early twentieth century, had spread the idea that space travel was not a mere fantasy. That led to the growth, in the 1920s and later, of what Alexander C. T. Geppert has called "astroculture": a "heterogeneous array of images and artifacts, media and practices that all aim to ascribe meaning to outer space while stirring both the individual and the collective imagination."[1] During the Cold War space race, spaceflight came to be embedded in national cultures of the superpowers, but also many other nations. Yet astroculture always encompassed more than spaceflight, real and imagined, because it drew upon older traditions and new contributions from astronomy, spirituality, and concepts of alien life. The ubiquity of space fiction series like

In order to make spaceflight real, one first had to imagine it. The rise of space science fiction in the nineteenth century, followed by nonfiction space advocacy in the early twentieth century, had spread the idea that space travel was not a mere fantasy.

Star Wars and *Star Trek* around the world today is a testament not only to the Americanization of global popular culture, but also to the influence that astroculture has had upon the imagination of people everywhere.

To analyze this disparate set of phenomena, I have divided this chapter into five brief sections: (1) the rise of "astrofuturism," which encompassed both factual and fictional advocacy for space travel as the future of the human race; (2) concepts of extraterrestrial life and their impact on space exploration; (3) the rise and decline of spaceflight enthusiasm during the space race and the cult of the astronaut; (4) spaceflight's impact on imagining Earth as a planet and its embeddedness in cosmic evolution; and (5) the rise of global astroculture through popular entertainment and the spread of spaceflight capability beyond the superpowers.

Space Science Fiction and Astrofuturism

Chapter 1 noted the important effect of early science fiction, particularly that by Jules Verne and H. G. Wells, on the imaginations of the pioneers of spaceflight theory. That genre emerged in the nineteenth century in the industrial countries of Europe and North America in large part because of technology's influence. If humans could do X, what could we not do? Ballooning, which began in

The ubiquity of space fiction series like *Star Wars* and *Star Trek* around the world today is a testament not only to the Americanization of global popular culture, but also to the influence that astroculture has had upon the imagination of people everywhere.

France in 1783, inspired many tales of buoyant flight to the Moon over the following century, before astronomy and ascents into the stratosphere demonstrated that the atmosphere had a limit—although popular understanding of that fact lagged decades behind the science. Verne, knowing the facts, proposed a cannon instead, although he failed to understand that instant acceleration would flatten his fictional passengers. Wells's *First Men in the Moon* (1901) and Kurd Lasswitz's important German spaceflight novel *Two Planets* (1897) used mysterious antigravity substances, a favorite device of early space fiction. Space travel had thus entered into the consciousness of at least the advanced, white-dominated part of the globe, even if actual flights seemed to be impossible or impossibly remote.[2]

The rocket began its conquest of fictional spaceflight in the 1920s, in part because of the worldwide echo of Robert Goddard's Smithsonian treatise *A Method of Reaching Extreme Altitudes* (1919). The works of Hermann Oberth and Konstantin Tsiolkovsky soon thereafter reinforced interest in the rocket as the way to spaceflight, especially in Central and Eastern Europe. That led to the production of the first realistic space films: *Aelita* (1924) and *Cosmic Voyage* (1936) in the Soviet Union, and *Woman in the Moon* (1929) in Germany. Beginning in the late 1920s, cheap, popular science fiction magazines and movie serials blossomed in the United States, featuring rocket pilots like

Buck Rogers and Flash Gordon. So prevalent was pulp science fiction in the 1930s that it damaged the credibility of spaceflight in the United States and tainted the word *rocket*. When the U.S. Army Air Corps offered funding for take-off assist rockets in 1938, a leading aeronautical engineer told Theodore von Kármán, a famous aerodynamicist at the California Institute of Technology, that he could have "the Buck Rogers job." When the Caltech group got increased army support in 1944, it called itself the Jet (not Rocket) Propulsion Laboratory.[3]

After World War II, V-2 and sounding rocket flights gave new legitimacy to the idea that spaceflight could be near, but the topic still had the lingering odor of a comic book fantasy. Advocates set out to sell Anglo American audiences, through fact and fiction, on the imminence and importance of space travel. Many of its leading figures were products of the interwar space societies—notably two former Germans, science writer Willy Ley and rocket engineer Wernher von Braun, and British author Arthur C. Clarke. Clarke was unusual in being equally successful at publishing fiction and nonfiction works about space travel, but science fiction writers like Robert Heinlein contributed to this advocacy as well.

Literary scholar DeWitt Douglas Kilgore has coined the term "astrofuturism" to describe this phenomenon, with its apogee in the 1950s. The central tenet was that the future of the human race lay in space; in fact,

So prevalent was pulp science fiction in the 1930s that it damaged the credibility of spaceflight in the United States and tainted the word *rocket*.

spaceflight was a guarantee of human progress because of the extraterrestrial knowledge, resources, and perspectives that it would open up. Central to the astrofuturist vision were tropes of global exploration, imperial conquest, technological utopianism, and (for the United States especially) the Western frontier. Notable breakthroughs in advocacy were Ley's 1949 book, *The Conquest of Space,* with leading space artist Chesley Bonestell, Clarke's *The Exploration of Space* (1951), and the *Collier's* magazine article series by von Braun, Ley, Bonestell and others (1952–1954). The series led to three spin-off books and three Walt Disney TV programs (1955–1957) featuring Ley and von Braun. Hollywood had already produced *Destination Moon*, a 1950 feature film based on a Heinlein book. It won an Academy Award for special effects. Still, what appeared in cinemas were mostly low-quality films like those in the horror and monster genres. Nonetheless, the astrofuturists succeeded in selling many in the English-speaking world and in Western Europe on the imminence of spaceflight even before Sputnik was launched.[4]

A parallel phenomenon took place in the Soviet Union in very different circumstances. After the war, leading rocket engineers like Sergei Korolev, Valentin Glushko, and Mikhail Tikhonravov, all secretly engaged in ballistic missile programs, began a campaign to legitimize space travel, which had been dropped for a decade due to Stalin's

Figure 9 German American rocket engineer Wernher von Braun stands in front of a Chesley Bonestell painting for *Collier's* magazine in 1952, holding a model of his booster for Walt Disney's 1955 space television programs. A leader in the German and U.S. army rocket programs, and later NASA, he also made himself into one of the most important popularizers of astrofuturism in 1950s America. *Source:* Smithsonian National Air and Space Museum (NASM 77-12796)

repression and World War II. Speaking to closed audiences, but also publishing in the Soviet press under pen names, they used the memory of Tsiolkovsky, who had died in 1935, to assert a Russian/Soviet first in the area of space travel. After Stalin died in 1953, and the Khrushchev cultural "thaw" began to open up the society, science fiction again began to flourish, and articles on spaceflight increased. In secret, Korolev and his associates also succeeded in selling satellite projects to the party leadership, leading to the Sputnik shock.[5]

The astrofuturist phenomenon did not end with the 1950s. Reinforced by the early space race, it became common currency in the 1960s. Spaceflight became a normal part of future visions; indeed many in the public, the press, and the political elite came to equate it with the human future. New advocates emerged, like the astronomer Carl Sagan, who became a public figure in the 1970s. Science fiction flourished, but also greatly diversified, taking much of it away from the optimism of its origins, as well as its gendered, white male character. The decline of the space race also undercut belief in spaceflight as the future. These phenomena will be explored below.

Alien Life

Ideas of extraterrestrial life were clearly not new when the spaceflight era arrived. The scientific revolution in early modern Europe meshed with Christian belief to create the expectation that the Moon and planets would be inhabited by other intelligent creatures. Otherwise, why would God have created those places? The "plurality of worlds" was widely accepted in seventeenth- and eighteenth-century literature. The rise of space science fiction in the nineteenth century provided another venue for imagining alien encounters.[6]

Growing astronomical knowledge in the late nineteenth and early twentieth centuries made much of our solar system seem uninhabitable, but it had the opposite impact on perceptions of Mars. After the Red Planet made a close approach in 1877, Italian astronomer Giovanni Schiaparelli claimed he saw straight "canali," a word that means channels but is usually translated as canals. After a wave of popular enthusiasm following another close approach in 1892, Martians became the subject of numerous science fiction stories like Wells's *The War of the Worlds* (1897) and Lasswitz's *Two Planets* (in which the Martians are benign). In the late nineteenth and early twentieth centuries, American astronomer Percival Lowell published books claiming the canals represented the handiwork of an advanced civilization. There were also repeated discussions of how to communicate with the Red Planet or, after the rise of radio, listen for their transmissions.[7] Cloud-shrouded Venus, closest to Earth in distance and size, also provided a subject for speculation. When the spaceflight movement arose between the world wars, its advocates naturally spoke of the power of rocket technology to explore these mysteries.

After World War II, the unidentified flying object (UFO) phenomenon added another dimension. It began in 1947 when a pilot flying over Washington state reported silver disks maneuvering near him, which the media soon dubbed "flying saucers." The many sightings that followed had precursors: several waves of mysterious airship sightings in Britain, the United States, and elsewhere between the 1880s to the 1910s seem to have been inspired by reports of aerial inventions and fear of them in enemy hands. In 1946–1947, Scandinavians and Northern Europeans claimed they saw "ghost rockets," another likely case of mass suggestion shaped by German V-2 testing over the Baltic in World War II and fear that the Soviets were doing the same. The UFO wave lasted much longer than any of the earlier ones, however, perhaps because it was a by-product of Cold War fear of, and fascination with, exotic weapons. The hypothesis that UFOs were extraterrestrial was not the only explanation—there was much early speculation about secret U.S. and Soviet aircraft, although that declined as the actual weapons development on both sides became better known. Another piece of evidence indicating that the Cold War was influential is that public interest in UFOs went into decline after that era ended. The extraterrestrial hypothesis never gained any traction with scientific, military, and political elites, and thus had no impact on national space programs, but its effect on the development of astroculture was nonetheless significant, through space fiction films, repeated media coverage, and the development of complex subcultures of believers.[8]

The actual exploration of the Moon and planets that began in the 1960s had critical impacts on public perceptions of alien life. U.S. Mariner spacecraft quickly undermined the already diminishing hope that larger life-forms, or even single-celled organisms, might exist on Venus and Mars. The former planet had temperatures hot enough to melt lead, while the latter looked like a cold, cratered desert. Extensive orbital photography in the 1970s made Mars more interesting again, because of the apparent role that floods had played in its distant past, but Viking's failure to detect any life in 1976 effectively ended U.S. Mars exploration for twenty years. Public opinion polls have demonstrated that extraterrestrial life has consistently been a strong motivator for public and political support for space exploration. Any early life discovery would have radically changed the course of space history. Instead, the negative result reinforced the decline of public interest and state funding for spaceflight that ensued after the Moon race.[9]

Nonetheless, popular fascination continued, leading to a shift in the perceived location of intelligent life to interstellar space and other star systems. In the early Cold War, when UFOs were perceived as extraterrestrial, the press and public often spoke of "Martians." As space exploration eliminated all hope for intelligent solar system life, or indeed anything beyond simple organisms, the focus of the UFO subculture and mainstream science fiction shifted to interstellar travel. It was not a new idea, but it soon became the main focus for representations of how humans would encounter aliens, notably in films—either we had that technological capability or they did, or both. Stanley Kubrick's *2001: A Space Odyssey* (1968) and Steven Spielberg's *Close Encounters of the Third Kind* (1977) and *E.T. the Extra-Terrestrial* (1982) showed benign aliens coming to Earth, whereas the TV and movie franchise *Star Trek* embodied the shift of human exploration to an imagined interstellar space. Darker visions of extraterrestrial life, like director Ridley Scott's *Alien* (1979) and its sequels, were the flip side of the same coin.

The Cold War expansion of astronomical tools and capabilities came to influence astroculture and scientific concepts of extraterrestrial life as well. The availability of large radio telescopes led American astronomer Frank Drake to propose listening to nearby stars for signs of alien civilizations. A new subfield, the search for extraterrestrial intelligence (SETI) grew out of his idea; Carl Sagan became one of its most persuasive advocates and popularizers, just as he was for the search for life in the solar system. His science fiction novel about SETI, *Contact* (1985), became a Hollywood movie in 1997 that brought the effort to a global public in a way that media coverage had not. Professional astronomy's search for evidence of other planetary systems had a more subtle influence, mostly after a technological breakthrough in the 1990s led to the actual detection of extrasolar planets. The subsequent discovery of thousands of systems, most notably by NASA's Kepler space telescope, has reinforced the message that there are a lot of possible locations for alien life.[10]

The search for life goes on and continues to shape public support for space programs and the broader astroculture. Astrobiology has revived the solar system search by emphasizing exotic life-forms on Earth and how they might apply to new locations, such as the subsurface oceans of the icy satellites of Jupiter and Saturn. And life was the scientific context when the U.S. Mars program resumed at the end of the 1990s with a "follow the water" strategy—although hope had receded to, at best, finding single-cell organisms underground or fossils of the same in the rocks. Recent movies have reflected these quests (e.g., *Europa Report* [2013] or *Life* [2017]), although interstellar aliens and settings remain dominant in popular entertainment, often with minimal regard for the laws of physics. Thus, regardless of the actual results of space missions and astronomy, aliens remain a pervasive subject of astroculture.

The Space Race and the Cult of the Astronaut

The first two Sputniks were already the subject of songs and poems, notably the doomed space dog Laika in Sputnik 2. But human spaceflight's quick, space race–driven emergence, combined with nationalism and the public's much greater identification with what humans would experience, meant that cultural production related to real space programs came to be almost entirely about spacefarers. The seven Mercury astronauts were instant heroes after NASA's announcement of their selection in April 1959; the cosmonauts had to wait for fame until their first flight, if they got one, such was obsessive Soviet secrecy. The two sides developed parallel cults; spacefarers were heroic, quintessentially masculine pilots, model citizens, and fathers, patriots, and true believers in their respective political systems. As international representatives, they were envoys of peace, notwithstanding that they were almost all (in the first groups) military officers. These cults were government sponsored, but they were also genuine for many people in the United States, the Soviet Union, and their allies.[11]

Such was the public success of the U.S. astronauts that the fighter/test pilot instantly became the image of the hero-astronaut in American movies and television, crowding out earlier representations of scientists, adventurers, and others. When scientists were aboard in movie plots after this time (and to a large extent also earlier), they tended to be obsessive, obstructive, incompetent, or evil. When NASA chose actual scientist-astronauts beginning in 1965, they found their role marginal, as the pilot-astronauts were completely in control of assignments. Only four got into space before the space shuttle era.[12]

In the USSR, the cosmonauts were heirs to previous pilot heroes, who had pulled off various feats for Stalin in the 1930s. Yet the state media emphasized their piloting role less, in part because the Communist Party frowned upon individualism, and in part because the Soviet cosmonauts were young, relatively inexperienced fighter pilots—Korolev and his engineers had designed fully automatic spacecraft in which the cosmonaut had little control. The pilot cosmonauts later had to carry out a behind-the-scenes struggle for more control as the vehicles became more complex. They also found themselves challenged by a new class of engineer-cosmonauts, often drawn from Korolev's design bureau.[13]

The cosmonauts who flew found themselves caught in a dilemma. Since in Soviet propaganda every mission was a success and everything went according to plan, and all spacecraft and rocket details were state secrets, they were forced to be evasive or dishonest in public, while being held up as models of "socialist honesty." This began with Gagarin, who had to cover up his harrowing reentry due to equipment failure, plus his ejection and landing on a parachute because the Vostok spacecraft could not slow down enough to be ridden safely to the ground. International aviation rules required that he had to land in his craft to get the record of first human spaceflight, so the Soviets ordered Gagarin to lie about it. He became a national and global hero, toured many countries, and gave many speeches for the party, yet he had to sustain all the lies he and the colleagues were trapped in. The strain may have contributed to his drinking and womanizing, quite in contrast with his public image.[14]

The string of propaganda successes from 1961 to 1965 did not prepare the Soviet public for the sudden deaths and disasters later in the decade, ending with the failure to match any of the lunar feats of the Apollo astronauts. Three deaths were a particular shock: Sergei Korolev's on the operating table in January 1966 at age fifty-nine (his obituary revealed that he was the heretofore anonymous but lionized Chief Designer); Vladimir Komarov's during his Soyuz 1 crash in April 1967; and Yuri Gagarin's in a training jet accident in March 1968. Regarding the latter death in particular, Soviet citizens circulated conspiracy theories and rumors in a society already awash in them because of the absence of an honest press. The widespread and genuine "cosmic enthusiasm" of the early 1960s, when it seemed that the USSR was first in the world and a utopian Communist society was not so far away, soured into the skepticism of the stagnant years of the late Soviet Union, and eventually into a Russian nationalist nostalgia about the good old days in space. As for the actual human space program, the shift to space stations in the 1970s and later simply could not sustain the same level of interest in the public, although it was episodically interesting. Like the U.S. astronauts in the shuttle era, cosmonauts were revered when encountered as individuals, but their names became almost unknown to the public.[15]

Disillusionment about the human space program in the United States set in during the late 1960s as well, thanks largely to urban riots, poverty, crime, and the Vietnam quagmire, leading to a growing resistance to spending tax money on space, and a growing distrust of technocratic elites and politicians. Apollo's glorious year of achievement from late 1968 to late 1969 temporarily reversed the decline in public support for NASA, but it did not alter the overall trajectory. Immediately after Apollo 11, opinion polls demonstrated weak support for more Moon landings, let alone ambitious programs like going to Mars. The conspiracy theory that the landings had been faked arose at this time, reflecting a belief in the federal government's dishonesty in the era of Vietnam and Watergate. The Nixon administration took its cue from public opinion and cut NASA's budget further, saving only the shuttle. The desperate agency switched to emphasizing the practical spin-offs of space research and its contribution to cleaning up the environment. In the 1980s, the early shuttle flights helped rebuild national pride and interest in spaceflight, but the astronauts remained anonymous beyond a few who made notable achievements like Sally Ride, who in 1983 became the first American woman in space.[16]

Gender, and also race, had always complicated the ultramasculine image of the hero-pilot astronaut or cosmonaut. NASA's rejection of women pilots in the early 1960s contrasted with Valentina Tereshkova's flight in 1963, making the United States look backward to feminists at home and abroad—although without significantly altering support for NASA as yet. The agency had another public relations problem in its all-white corps. A potential black candidate the Kennedy administration pushed failed to progress in air force test pilot school due to either his limitations or discrimination, depending on who one listened to. The lack of black astronauts contributed to an especially steep decline in space enthusiasm in the African American community in the late 1960s, although the primary motivation was the feeling that money was being squandered on the Moon when it could have been spent on the impoverished and the inner cities—a feeling shared by most white liberals and leftists.[17]

When the shuttle program allowed NASA to reopen astronaut recruitment in the mid-1970s, it was in a different political, cultural, and legal environment. Gender and race discrimination were now formally illegal, resulting in the selection of the first women and blacks in 1978. Nonetheless, the media reaction to the first female astronauts was often insensitive and sexist, and the male engineers at NASA were often a little clueless.[18]

On the other side, Tereshkova's flight inspired a lot of women and girls in the Eastern Bloc, but the propaganda about how it demonstrated women's equality in socialist societies was undercut by no more female flights for nineteen years. The Soviets finally orbited Svetlana Savitskaya in 1982 with the transparent objective of preventing Sally Ride from becoming the second woman in space. Over time the American and Soviet/Russian shuttle and station programs normalized the gender and racial diversity of crews (in part by also launching crewmembers from many other nations), but to this day it has not fully displaced the paradigm of the macho pilot-astronaut in popular culture.[19] The public identification of human spaceflight with "the space program" also remains pervasive, such that when the shuttle program ended in 2011, many Americans believed that NASA had closed.

"Planetization" and "Cosmic Evolution"

The most profound impact of spaceflight on consciousness and culture, and perhaps the hardest to demonstrate, is how space images and scientific discoveries have shaped perception of our planet and our place in the universe. What little scholarship there is on the topic has been about Apollo photos of Earth and the shaping and dissemination of images from the Hubble Space Telescope. A related question,

The public identification of human spaceflight with "the space program" also remains pervasive, such that when the shuttle program ended in 2011, many Americans believed that NASA had closed.

but one that is difficult to answer, is how much space images and knowledge have altered human perception of its purpose and place in the universe—questions normally in the domain of religion and philosophy.

Much has been attributed to two Apollo images that have acquired informal names because of their ubiquity:

Figure 10 On December 24, 1968, during the first human mission to the Moon, Apollo 8 astronaut William Anders took the most influential space image of all time. Dubbed *Earthrise* soon after, it has become ubiquitous as it encapsulates for many the beauty and fragility of Earth isolated in a vast universe. It and other Apollo images have been central to the "planetization" of human identity in the Space Age. *Source:* NASA

Earthrise, taken by the Apollo 8 astronauts over the Moon in December 1968, and *Full Earth* (also called *Blue Marble*), an Apollo 17 photograph from after launch in December 1972 showing a completely illuminated day side with Africa and Arabia centered. These were by no means the first pictures of the planet in space, nor the first imaginings. Greco-Roman, Renaissance, and early modern writers described what it might look like. Popular astronomy books included Earth illustrations as early as the nineteenth century. Sweeping views taken by stratospheric balloons, sounding rockets, and early human missions—showing clouds, landscapes, and a curved horizon—fascinated many. In 1966, Lunar Orbiter 1 took the first *Earthrise* photo and a NASA applications technology satellite made a near *Full Earth* image from geostationary orbit. Yet there is no denying the unique impact of Apollo's sharp, full color images, as opposed to fuzzy black-and-white; that humans, not robots, took them added to their appeal. *Earthrise* and *Full Earth* not only became media icons, they also became common in the signs, banners, and advertisements of environmental movements in the 1970s and later. In fact, the Western media has often simplistically attributed the rise of modern environmentalism to the Apollo images of a fragile, borderless Earth, ignoring those movements' origins and earlier history. Yet the very ubiquity of the two images is proof that they have somehow altered astroculture and human perception.[20]

Years before Sputnik, French philosopher Pierre Teilhard de Chardin called the process of coming to terms with our home being just one planet in space "planetization." I agree with Geppert that this concept is usefully distinguished from globalization.[21] The integration of world cultures, economies, and governmental systems shapes human perceptions of the planet too—but largely as questions of identity as members of ethnic and national groups versus a common humanity. Seeing Earth in the void, embedded in an unfathomable universe and providing the only place we know can support life, at least life as we know it so far, creates an emotional experience distinctively different than imagining the globe as an interconnected set of human societies, for good or ill.

A related question is the impact of the growing knowledge of the vast distances, timescale, and evolution of the universe, a question often intertwined with discussions of the probability and nature of extraterrestrial life. Carl Sagan, in his popular works and many television appearances in the 1970s, 1980s, and 1990s, was particularly influential in his attempt to create an awareness of our place in the universe and how it relates to the past and future of the human race. Rooted as he was in astronomy, robotic planetary exploration, and the network of scientists interested in SETI, he presented spaceflight in a very different context than astrofuturists of the first generation did. They wanted to sell space travel as the desirable and immediate future of humanity; Sagan came into prominence as a popularizer after the space race had lost its momentum and human spaceflight some of its luster. Obsessed with extraterrestrial life since childhood, and having worked on the chemical and biological origins of life, as well as in mainstream planetary science, he presented the human achievement of spaceflight in the context of billions of years of "cosmic evolution." That concept had arisen in the mid-twentieth century when scientists and philosophers linked Big Bang cosmology (which empirical astronomical evidence validated by the 1960s) with nebular theories of stellar and planetary formation, origins-of-life theory, and Darwinian evolution into one gigantic, naturalistic (and implicitly atheistic) explanation for the history of everything. Sagan successfully sold this grand perspective to millions of readers and TV watchers in the United States, the English-speaking world, and beyond.[22]

Does this mean cosmic evolution became a factor in how most people imagined their place in the universe? Given the weak grasp of science even in the privileged parts of the world, and the durability of traditional religions everywhere, it seems quite unlikely, and in any case, the scholarship scarcely exists to prove such a hypothesis one way or another. Yet Sagan's influence on astroculture and human self-perception cannot be lightly dismissed, given his popularity, nor can that of other science writers and scientists who followed him, like Neil deGrasse Tyson.

Once the Hubble Space Telescope was repaired in 1993, its images added another dimension to the public imagining of the universe. Scientists at the NASA-funded Space Telescope Science Institute responded to intense public interest by founding the Hubble Heritage Project, to process images that were likely to be of particular interest to the nonscientific public. The team has used techniques like false color and aesthetic considerations to make choices about how to display the digital data. Their influences appear to be romantic paintings and photographs of frontier landscapes, notably in the American West.[23] The project manipulated images to make particular scientific points while impressing the public with the beauty of the universe. NASA has extended this process by combining images from the different Great Observatories, superimposing X-ray or infrared data of the same objects in order to make scientific content more engaging and understandable. The appetite for such images, as evidenced by the popularity of space telescope pictures in the media, online, and in books, appears to indicate that these efforts have reached a receptive public and transcend their public relations purpose, which is to sell NASA programs.

Do these rather isolated, overwhelmingly American examples demonstrate that humanity has been "planetized" and achieved a consciousness of cosmic evolution? Hardly. Yet they do demonstrate that space images have shaped astroculture and the public imagination at least in the English-speaking world, and likely in much of the rest of the globe as well. Beyond that, much more scholarship is needed.

Globalizing Astroculture

Describing the global spread of a heterogeneous and multifaceted astroculture is even more challenging, given how little has been written about anything but the United States, the Soviet Union, and Europe. It is possible to discuss very briefly, however, the worldwide diffusion of space advocacy, spaceflight imagery, and science fiction entertainment.

Early science fiction and astrofuturism was an almost exclusively Euro-American male phenomenon, in which existing social, racial, and gender hierarchies were often taken for granted. The future "conquest of space" was set within a conventionally imagined history of European global exploration, Western settlement, and technological superiority. The early space movement was overwhelmingly concentrated in Russia/USSR, Austria, Germany, France, Britain, and the United States. After World War II, Western European societies organized the first International Astronautical Congress (1950) and International Astronautical Federation (1951), and then quickly integrated U.S. and Soviet representation. But in the wake of

Sputnik, advocacy societies sprang up around the world, often in conjunction with modest and sometimes extensive space programs. The superpowers sent space vehicles and spacefarers, as well as glossy publications, films, and exhibits, to every continent as part of their global struggle for influence. Latin Americans, Africans, and Asians turned out in droves for spacecraft and astronaut/cosmonaut tours; they often identified space accomplishments as human, rather than just as those of one country or Cold War bloc.[24]

What that means for astroculture in those countries and continents needs study. Just as a thought exercise, I would like to contrast the circumstances surrounding the dissemination of spaceflight ideas and images in Japan and China, two Asian countries that quickly became significant space powers. Postwar democratic, capitalist Japan integrated into the West and built its space program partly in collaboration with the United States and partly through its own institutions, leading to a human space program that used American and Russian rockets to transport its astronauts. Japan's open culture imported European and American space advocacy literature, science fiction, and television and movie space representations, and produced its own. Communist China, by contrast, developed a secretive military space program on the Soviet model, but it opened its society after the 1980s to a partially capitalist economy and a controlled, but still freer flow of information, tourism, and entertainment across its boundaries. To build scientific and technological capability and signal that strength internationally, the government created a human space program and later a robotic Moon and Mars program. Government space propaganda and information has remained central to Chinese astroculture in a society still tightly controlled by the party. But after China orbited its first astronaut in October 2003, he became a national celebrity in a new media culture that broke the bounds of the old Communist hero role model.[25] The economic and cultural opening also allowed Western space science fiction, notably movies like the *Star Trek* series, into the Chinese market, just as they had earlier entered the market in Japan. As different as the two societies are still, the Americanization and globalization of entertainment and popular culture thus produced a degree of convergence in the kinds of spaceflight images reaching their respective publics.

The global reach of Hollywood motion pictures and television incorporating space themes implies that they are now one of the most important means by which spaceflight images reach a world public, followed by news and popular media coverage of the events or image products of actual space missions. That by no means implies that astroculture has become globally unified. Given that even in one country it is a "heterogeneous array of images and artifacts, media and practices that all aim to ascribe meaning to outer space," to quote Geppert again, such an outcome is scarcely imaginable. What one can say is that, whereas cultural representations of spaceflight, real and fictional, were once almost exclusively confined to advanced Euro-American societies, they are now a global phenomenon.

Conclusions

From its origins in science fiction and early space advocacy, astroculture has grown into a panoply of national and transnational products, discourses, and genres (perhaps it should be plural, not singular: astrocultures). Among its most noteworthy features are an astrofuturist belief that spaceflight represents the future of the human race (an optimism that eroded, but did not vanish, after the Moon race), that it would lead to encounters with alien life (a belief that seems undimmed by the failure so far to find any), and that spacefarers are heroes (despite the growing routineness of their flights). Space images, notably of Earth, have become embedded in global culture in such a way that they may have contributed to the planetization of human identity and an understanding of our species' place in cosmic evolution.

All of these factors have contributed to sustaining public support for government and private space exploration and exploitation after the end of the Cold War, notably in human spaceflight. But the decline of international competition as the driving force has also demonstrated the limits of astroculture's influence. The human-dominated solar system expected by astrofuturists and science fiction writers has failed to materialize, even as robots have reached every planet. Yet the dream seems not to have died, which perhaps helps explain why human voyages outward remain on the agenda for the 2020s and beyond.

6

HUMAN SPACEFLIGHT AFTER THE COLD WAR

On July 20, 1989, the twentieth anniversary of Apollo 11's landing, President George H. W. Bush announced on the steps of the National Air and Space Museum that American astronauts were going back to the Moon and on to Mars. On January 14, 2004, his son, President George W. Bush, made a very similar announcement at NASA Headquarters.[1] Little came of either declaration. Instead, human spaceflight remained firmly stuck in low Earth orbit. The major space powers focused on keeping alive or completing the projects of the 1970s and 1980s: the U.S. Space Shuttle, the Russian Soyuz spacecraft and the Mir station, and what became the International Space Station.

The two new players were the Chinese and private investors. One year after the former orbited their first astronaut, a privately funded rocket plane won a prize for the first nongovernment vehicle to fly twice in two weeks above 100 km (62.1 miles), the widely accepted definition of where space begins. Suborbital space tourism seemed to be on the verge of reality, yet no tourists have flown to date, other than a few multimillionaires who paid the Russians for seats on the Soyuz craft that took crews to the ISS. There are many signs in the late 2010s that both suborbital space tourism and human Moon flights will soon be attempted, but the three decades after 1989 demonstrated that, without the driving force of a space race, human spaceflight limped along, sustained mostly by geopolitical signaling and the need to maintain the jobs and infrastructure constructed after Sputnik.

The Space Shuttle and Space Stations

The January 1986 *Challenger* accident fundamentally shaped the second act of the shuttle program. No longer would it launch commercial satellites; classified national security missions also stopped after shuttles orbited the few payloads that could not be put on expendable launch vehicles. Shuttle flights did not resume until September 1988 and the last U.S. Defense Department mission was flown in 1992.[2] A heterogeneous collection of NASA flights came to dominate the manifest—the Hubble Space Telescope and its repair and maintenance missions, planetary and other science spacecraft left over from before the accident, and Spacelab missions in which crews, often including European, Canadian, or Japanese astronauts, spent about two weeks doing science in orbit. But it was flights to space stations that ultimately became much of the shuttle's flight schedule and raison d'être—a return to the purpose for which it was originally conceived.

NASA's space station *Freedom*, which had originated as a Cold War project of the Reagan administration, was designed to be assembled out of modules lofted by the shuttle. But *Freedom* was troubled from the outset. In one of the more egregious cases of underpricing to get "buy-in" from the political establishment, a familiar behavior pattern in the U.S. military-industrial complex, NASA had promised in 1983–1984 that it could build a very large, multipurpose station by 1992 for only $8 billion. It then succeeded in getting the European Space Agency and Japan to promise to add their own laboratory modules, and Canada would supply a manipulator arm derived from its shuttle version. Yet by the original target date, the agency and its contractors had built virtually no flight hardware. Instead, billions had been spent on multiple paper redesigns, exacerbated by a weak, diffuse management structure reflecting NASA's culture of dividing the spoils among competing centers. While the agency framed the program as a stepping-stone to deep space exploration, it acted as if its first priority was keeping the large ground infrastructure built for Apollo in business—an objective shared by congressmen and senators who represented districts and states with NASA facilities or major contractors.[3]

Nevertheless, the endless budget increases and redesigns undermined the station's political viability, adding to NASA's troubles at the end of the Cold War. President Bush's 1989 speech launched the Space Exploration Initiative, which aimed to complete *Freedom*, build a Moon base, and put humans on Mars by 2019. But the grand program died a rapid death in Congress when NASA came out with a politically intolerable estimate of half a trillion dollars ($1 trillion today)—just as the space race rationale was evaporating.[4] Then came the Hubble mirror embarrassment in mid-1990, followed by foreign and domestic problems that weighed on the national budget. The large increases NASA had received in the late 1980s came to a sudden halt. As I noted in chapter 3, the Bush administration, frustrated over the agency's poor performance and bureaucratic ways, fired the NASA Administrator Richard Truly and installed an industry outsider, Daniel Goldin, in April 1992. His objective was to shake up the agency with "faster, better, cheaper" methods adopted from military space programs. But because the human spaceflight centers needed the space station to sustain their existing workforces and provide a program beyond the shuttle, he found himself trying to save a project that was the epitome of the sluggish, self-interested NASA he was trying to reform.

Freedom's salvation came, ironically, by merging with the former enemy's program. The Soviets had orbited Mir in 1986, a more elaborate and flexible evolution of the civilian Salyut stations. It incorporated a multiple docking adapter that allowed several more modules to be added over the next few years, enabling longer cosmonaut stays and more sophisticated scientific experiments. But the USSR's collapse in 1991 produced severe budget cuts for all government departments, including the human space program. A Soviet version of the shuttle called the *Buran* (Snowstorm) was canceled after one unmanned flight in 1988, as was the Saturn V class Energia (Energy) booster that launched only it and an ill-fated 1987 test of a laser battle station. A replacement for the aging Mir was put off and the Russian space agency Roscosmos, founded in 1992 to give Western organizations a partner, had little money and no authority over the military space forces or the powerful design bureaus being privatized into companies.

In 1993, the new Clinton administration was seriously concerned that Russian rocket engineers, who were being thrust into poverty for lack of pay, would work for Iran, Iraq, North Korea, or other nations seeking a ballistic missile capability. Moreover, *Freedom* was very close to cancellation in Congress, surviving by only one vote in June. In September, Vice President Al Gore and Prime Minister Viktor Chernomyrdin agreed to merge their station programs. Russian modules would be integrated with the U.S., European, Japanese, and Canadian components, with the whole retitled the International Space Station. NASA would buy some Russian components for the initial station, transferring desperately needed money into their industry. As preparation for the joint station, space shuttles flew ten missions to Mir, for which the Russians also received NASA support. A few cosmonauts rode up on the shuttle, while American astronauts made long-duration missions with their Russian counterparts on Mir, sometimes riding the Soyuz up or down. Shannon Lucid set a new American record of 188 days in space, although that was far exceeded by Valeri Polyakov, who heroically spent 437 days and 18 hours in orbit, still the single-flight record.[5]

By the late 1990s, Mir become increasingly dilapidated and unsafe. One NASA astronaut and his Russian colleagues experienced a serious fire emergency; another was aboard when a poorly conceived experiment in cosmonaut manual docking of a Progress supply ship led to a collision and the rapid decompression of one of the scientific modules. American visits and stays continued until 1998 nonetheless. Soon after the U.S. Space Shuttle–Mir program ended, the Roscosmos chief announced that it would be retired for lack of money and because NASA was pressing Russia to shift its full attention to the ISS. The last regular Russian mission was in 1999, but the station had a strange afterlife when the Energia company, the former

Figure 11 ISS astronauts Oleg Novikskiy, Fyodor Yurchikhin, Jack Fischer, and Peggy Whitson (station commander) share a meal in April 2017. Russian-American collaboration has been critical to the International Space Station project since 1993, regardless of the uneven state of relations between the two countries. *Source:* NASA (iss051e020883)

Korolev design bureau, made an agreement with a private U.S.-based group, MirCorp, to continue it as a for-profit station. The consortium funded a two-cosmonaut mission in 2000 to check it out and try to repair it. But the United States pressured Russia to get rid of this distraction from the ISS. In March 2001, Russian mission control directed Mir to burn up in the atmosphere over the empty South Pacific.[6]

NASA had reorganized its part of ISS and began to produce actual hardware, but it had new problems: an underfunded Russian space industry that could not deliver anything on time. In particular, the *Zvezda* (meaning "star") control module, which had originated as the Mir-2 core section, was late by a couple of years, delaying station assembly. In December 1998, a shuttle crew first linked two modules, *Zarya* (meaning "dawn"), paid for by the United States, but Russian-built and launched, and *Unity*, a U.S.-built docking node. But then a nineteen-month hiatus ensued for *Zvezda*. After it was launched in July 2000 and tested for habitability, Russia launched Expedition 1 on a Soyuz in November of that year, with two Russian cosmonauts and an American station commander. That began a period of continuous human presence in space that has not ended to this day.[7]

Constructing the ISS became the shuttle's primary mission until it stopped flying in 2011. Indeed, it was the shuttle's salvation after a second fatal accident in early 2003 again killed seven astronauts—*Columbia* broke up on reentry at the end of a non-ISS science mission that also carried Israel's first, and so far only, spacefarer. The cause, like that of *Challenger* in 1986, was during the launch phase, but unlike that accident, the disaster did not manifest itself until entry into the atmosphere. (*Columbia* was struck by a block of foam off the external propellant tank that smashed a hole in the leading edge of the wing, whereas one of *Challenger*'s solid rocket boosters had failed.) The second disaster demonstrated that

NASA still had a problem with its safety culture, because repeated warnings and nonfatal incidents had been inadequately addressed. Both failures also demonstrated that the fundamental shuttle design, which included no launch escape system, made it the most dangerous U.S. human spacecraft ever built. But any discussions of ending the program went nowhere because to retreat from human spaceflight would undercut America's superpower status; all non-Russian ISS modules were designed for the shuttle payload bay in any case. That led to George W. Bush's 2004 announcement that the shuttle would retire once the station was complete. Emphasis would shift to a new Moon-Mars program that would land astronauts on the lunar surface by 2019—or so it was hoped.[8]

The two-year gap in shuttle flights forced the ISS partners to downsize to two-person crews commuting on Soyuz spacecraft and supplied only by Progress vehicles (a robotic Soyuz in which the orbital module carried cargo and the return capsule was replaced by propellant tanks for refueling the station). Once U.S. launches resumed in 2005, however, the station quickly expanded to its massive planned size, 356 ft. (109 m) long and 239 feet (73 m) wide across four double solar electric power arrays generating 84 kilowatts of power, with a total Earth mass of 925,000 lbs (420,000 kg). The station is normally crewed by three Russians, two Americans, and one astronaut from Europe, Canada, or Japan. It is, and probably will be for a very long time, the largest human object ever put into space. One of the most remarkable things about the ISS is simply that it works, despite its multinational production. There were assembly problems and crises, notably one torn solar array that required risky astronaut spacewalks to fix, but all the modules and components functioned well together, a triumph of international project management.

Although the ISS was a technical success, the costs were huge and the process of getting there was plagued with waste and delays. The entire station project so far is estimated at around $150 billion, an expenditure on the order of the inflation-adjusted price for the Apollo program (which cost $25 billion in 1960s dollars). Roughly three-quarters of that was spent by the United States, notably when one includes the shuttle launches to the station at a price tag of over a billion dollars each. (So much for the shuttle's founding claim of making spaceflight cheaper.)

Is the ISS worth that huge expense? Not enough time has passed to make a full judgment, but there are some clear pros and cons. As an exercise in developing a multinational human space project that could be a model for deeper trips into the solar system, the ISS has been a sterling success. As a means to understanding the physiological effects of long-duration spaceflight necessary for going to the Moon or Mars, it has been valuable, although if the partners had built a smaller, purpose-built station, perhaps with three instead of six crew, it would have cost much less. As a science platform, the completed ISS hosts a wide variety of experiments in medicine, zero-gravity materials processing, Earth observation and even particle physics (the $2 billion Alpha Magnetic Spectrometer was attached to the station to look for dark matter, but it has nothing to do with the astronauts inside). If science were the only rationale for ISS, however, the results are too meager and the costs much too high to justify building it.

As always with human spaceflight, the ISS is more about global and domestic politics than science. While it left American astronauts stuck in low Earth orbit, because NASA had no budget to do anything else, it sustained the agency's infrastructure and capability after the end of the space race and helped confirm America's status as the largest space power. It provided a lifeline to survival for the Russian program, boosting Russia's claim to still being a major nation, and it was a mechanism for Europe, Canada, and Japan to build human space capability and experience without having to develop their own crew launch and recovery systems.

As for the shuttle program, it did finally come to an end in mid-2011, after 135 missions, once it had delivered the last major U.S., European, and Japanese components, plus strategic spare parts. For a significant fraction of the American public, the end of the shuttle meant the end of NASA and Americans in space—to them, the shuttle program was NASA. Expeditions to the ISS were and are too low profile for most, particularly when they were launched on Russian rockets. Yet the station occupancy continues to this day, and is scheduled to go on until at least 2024—the current lifetime the major partners have promised to support.

New Players and New Programs

Even as the major space agencies struggled to complete the ISS, several new initiatives in human spaceflight began to take shape after 2000: in China, in private space tourism, in NASA-funded human spaceflight vehicles, and in American space corporations that arose to challenge the military-industrial enterprises from the missile and space races. By the late 2010s, a new era appeared to be dawning in human spaceflight, with results that are far from clear.

One thing did become apparent: the People's Republic of China's determination to become a major space power in every dimension of spaceflight. Orbiting Yang Liwei in *Shenzhou 5* (*shenzhou* meaning "divine vessel") on October 15, 2003, made China only the third country to put an astronaut in space with its own launch vehicle and spacecraft. The objectives of Program 921, as it was bureaucratically titled, were (and are) to build a space station in order to cement China's claim to be a global power by signaling its national science and technology capability. Fostering national pride and public loyalty to the party leadership were not part of the original objectives, but were bonuses of successful missions.[9]

Program 921 began in 1992, after a premature attempt to start a human spaceflight project in the 1970s. The new project grew out of leader Deng Xiaoping's reform efforts to make the nation and its economy competitive with the West. It was and is run by the military, but like Soviet programs, in a symbiosis with civilian agencies, as there are no internal or external reasons to make a clear distinction between civil and military space efforts. Program administrators seized the opportunity created by the USSR's collapse (an event that shocked the Chinese Communist leadership and confirmed the need for reform) by licensing some Russian space technology. The Shenzhou spacecraft looks very much like Soyuz, but enlarged and upgraded. The reentry module also carries a crew of three, but in less cramped conditions, and the orbital module (additional living and experiment quarters) was replaced by a Chinese module with solar panels and systems that allow it to fly independently.

The pace of this program has been very measured. Shenzhou missions began in November 1999 with the first unpiloted test flight, followed by three more up to 2002. After Yang's one-day orbital journey, the next mission was not for two years, with two crewmembers, followed by a three-year interlude before *Shenzhou 7* flew with three crewmembers, two of whom made a spacewalk. (Their spacesuits were also based on Russian designs.) It was exactly the U.S.-Soviet steps of 1961–1965, but done without either the urgency or multiple missions of that race. Another three years passed before an uncrewed *Shenzhou 8* docked with the small orbital station, Tiangong-1 (Heavenly Palace 1), followed by a crew that did the same thing a year later, with one astronaut being the first Chinese woman in space. The most recent mission, as of this writing, was a thirty-three-day stay in fall 2016 by three crewmembers on Tiangong-2.

The China National Space Administration, founded in 1993 as an agency of the civilian defense procurement bureaucracy, has announced a program for building a Mir-sized, modular station, beginning around 2020, thus fulfilling 921's original objective. Due to technology export control and theft concerns, the U.S. government has so far excluded active cooperation between NASA and China, making its program entirely separate from the ISS and other multinational, American-led projects. One thing that does appear certain, however, is that China will build its human program regardless of what happens with international cooperation.

As Program 921 unfolded, so did the first attempts to create private human spaceflight projects and spacecraft. The ill-fated attempt to revive Mir was rooted in the frustrations of American space enthusiasts who grew up in the 1960s and wondered why the promised future of Moon bases and Mars missions never happened. Often they were highly critical of NASA and were adherents of ultracapitalist, libertarian politics. One of them was space entrepreneur Peter Diamandis, who created the X-Prize, modeled on the aviation prizes between the world wars that had spurred the growth of flight technology. Scrapping together donations and a risky bet on the insurance market, he offered a $10 million prize for the first non-government piloted spacecraft capable of carrying three persons to make two flights in two weeks above 100 km (62.1 miles). The objective was to stimulate private space tourism at the lowest level of difficulty—still not by any means easy—by lofting passengers on suborbital trajectories just high enough for them to claim the status of space travelers.

The X-Prize spurred a flurry of rocket development activity all over the world, but only one group had a serious chance to win: famed aviation designer Burt Rutan's Scaled Composites aerospace company, with funding from Microsoft billionaire Paul Allen. Rutan designed SpaceShipOne, a spaceplane with a unique folding tail that during the descent slowed the craft down like a badminton shuttlecock. Dropped from a large, Rutan-designed carrier aircraft, SpaceShipOne made three flights from Southern California that passed the 100 km line, the last two winning the prize in fall 2004. Each flight was made by one of

Figure 12 SpaceShipOne and its carrier aircraft White Knight in August 2005, upon delivery of the former to the National Air and Space Museum. The spaceplane's team won the Ansari X-Prize in 2004. It seemed to herald the imminent arrival of suborbital space tourism, yet it proved much harder than expected to begin actually carrying passengers. Photo by Eric Long. *Source:* Smithsonian National Air and Space Museum (NASM 2005-24511)

two company test pilots, flying with ballast to represent the two passengers.[10]

Simultaneously, orbital tourism emerged because the Russians decided to sell occasional Soyuz seats to the ISS There were precedents: an Englishwoman had flown to Mir in 1991, although the private funding from Britain had fallen through, and in 1998, a Japanese media company had paid to send one of its reporters to Mir. In April 2001, American investor Dennis Tito became the first to use his own money to fly into space, making a visit to the ISS over the objections of a reluctant NASA. Only six others did it up to 2009, one of them twice, after which all Soyuz seats were needed to crew the station. These orbital tourism trips cost $20–$40 million dollars and were confined to a handful of physically fit, ultra-rich people who had months to train at Russian centers. It was barely something that could be called tourism. The opportunity may reappear. Roscosmos has announced that it will be cutting back to two Russian crewmembers on the station, and NASA will soon be flying its astronauts again on American spacecraft.

As for the X-Prize, Rutan and Allen's 2004 win seemed to herald the imminent arrival of suborbital space tourism. The founder of the Virgin corporate empire, Richard Branson, funded Scaled Composites' SpaceShipTwo, which could carry two pilots and six passengers. His new company, Virgin Galactic, also built an elaborate spaceport in the New Mexico desert for its "astronauts." Despite a ticket price of around a quarter of a million dollars, hundreds put down deposits or full payments. Yet a series of technical challenges, and two fatal accidents, have resulted in no tourist flights as of this writing. The first SpaceShipTwo broke up on a powered test flight in October 2014, killing

These orbital tourism trips cost $20–$40 million and were confined to a handful of physically fit, ultra-rich people who had months to train at Russian centers.

one pilot and seriously injuring the other. The development costs have run far beyond what Branson expected, and the process confirmed what government-funded human spaceflight programs had painfully learned, namely, building systems to loft people to great altitudes and thousands of miles per hour is expensive and risky. Trying to insure at least reasonable safety costs a lot of money. Moreover, if there is ever a fatal accident with paying tourists, the market for seats might collapse and government regulation would certainly become tighter.

The development cost alone has caused several space tourist ventures to die from lack of venture capital. Other than Virgin Galactic, only one survives, Amazon.com billionaire Jeff Bezos's New Shepard, precisely because he does not need anyone else's money. New Shepard has a reusable booster and an automated capsule for six passengers that returns by parachute. Built by his space company, Blue Origin, it has been tested in West Texas since 2015. Thus, if suborbital space tourism finally arrives around 2020, very much later than expected, it will remain firmly in the zone of adventures for the rich, although the passengers will not have to be as well-heeled as the first orbital tourists.

Blue Origin was the first of two major space firms launched by Internet entrepreneurs with spaceflight ambitions. Elon Musk, who had made a fortune with the PayPal online payment service, created SpaceX in 2002 to work toward his imagined future as a founder of a Mars colony. Frustrated by the prices charged by the traditional aerospace firms, he set out to create a vertically integrated rocket company that made its own engines and most everything else. SpaceX's first vehicle, a small satellite launcher called Falcon 1, was not very successful, failing the first three times it was launched from 2006 to 2008. These failures drove the firm close to bankruptcy. After two successful launches, Musk decided to give up on the small satellite market and make a bid for the geostationary and military markets with the Falcon 9, which grouped nine of his Merlin engines into one vehicle.[11]

Falcon 9 was already under development because NASA had given SpaceX development money in 2004, leading to a contract in 2006 to supply cargo to the ISS. Musk, inspired by Apollo (as was Bezos) and obsessed with human spaceflight, used that contract as way to develop his own crewed spacecraft, Dragon. For the cargo version, the pressurized return capsule would carry racks instead of seats and instruments. Unpressurized cargo could be put into a "trunk" on the attached propulsion module.

The NASA commercial cargo project had arisen because the shuttle was phasing out and was far too expensive to be a carrier. Michael Griffin, who became NASA Administrator in 2005 to implement George W. Bush's Moon-Mars program, hoped to save money through the innovative use of commercial resupply to the ISS, rather than depend on costly Russian, European, and Japanese vehicles. To foster competition and reduce vulnerability to launch failures by one rocket type, NASA also let another space station supply contract that Orbital Sciences, a pathfinder commercial spaceflight company from the 1980s, took over in 2008. SpaceX's Dragon and Orbital's Cygnus completed test flights to the ISS in 2012 and 2013, respectively, initiating the era of U.S.-launched ISS cargo resupply.[12]

By then, NASA was funding commercial crew transports to the ISS. In 2009, the new Obama administration approved their development to shorten the post-shuttle period of dependence on the Russian Soyuz. After several rounds of competition, hampered by budget cuts from an unenthused Congress, NASA announced in 2014 that it was giving two companies contracts: SpaceX for its Crew Dragon and aerospace giant Boeing for its CST-100 capsule.[13] Both were supposed to fly with astronauts by 2017—and neither has. Once again, building relatively safe human spacecraft has proven more difficult and expensive than was hoped.

As all this was happening, the Bush administration's Moon-Mars program was running aground. When Griffin took over NASA in 2005, he had focused the newly named Constellation Program on returning astronauts to the Moon and creating a lunar base in the 2020s. That seemed more feasible in the nearer term than a Mars expedition, for which the lunar enterprise was to build technology and experience. But Bush administration funding was never realistic, as it assumed the shuttle would complete the station and retire sooner than it did, freeing up funds to pay for the new program. Neither Congress nor the administration was willing to increase NASA's allocation to cover the shortfall, as there were many other national priorities and no sense of urgency about the project. It demonstrated that once again, just as with the first Bush pronouncement in 1989, that trying to replicate Kennedy's Apollo speech was doomed when there was no perceived crisis.[14]

As a result, Constellation was seriously underfunded by the time Barack Obama came into office, with the first landing slipping into the 2020s, in part because there was no money to develop the lunar landing vehicle. The new president moved to cancel Constellation in 2010, which went over poorly in Congress and the aerospace industry. Senators from states with a strong vested interest in jobs created by the Orion crew vehicle, which looked like enlarged Apollo command and service modules with solar panels and a crew of four, and by the boosters, which were derived from space shuttle propulsion and tank elements. With support from industry lobbyists and other congressmen, the senators forced a compromise. Orion would survive, as would a Saturn V–sized heavy booster called the Space Launch System (SLS)—wags called it the

Senate Launch System, because it seemed as if it had been designed on Capitol Hill.

Not surprisingly, Orion/SLS development soon dropped behind schedule, as ultimately it had to fit into a NASA budget still funding ISS flights plus commercial cargo and crew vehicle development, plus the James Webb Space Telescope, a larger, infrared-focused successor to the Hubble that ran way over budget. NASA's new announced objective for Orion was to rendezvous with an asteroid, but then that was reduced to sending a robotic spacecraft to pick a boulder off an asteroid and bring it to a distant lunar orbit, where the astronauts would sample it. This mission attracted little support in Congress or the scientific community and was effectively dead by the end of the second Obama term. All of this activity, plus gathering long-duration spaceflight experience on the ISS, was to be part of NASA's "Journey to Mars," with expeditions in the 2030s. That did not look very convincing either.

Nonetheless, Orion/SLS has consumed several billion dollars per year while moving at a pace that makes the Chinese look they are in a hurry. The agency launched one Orion command module in 2014 on a reentry test flight, using a commercial booster, but that is the entire flight history of the program so far. The first flight of the SLS, which is to send an unpiloted Orion into lunar orbit, has slipped into late 2019, as of this writing. The first human mission, with an objective of several weeks in lunar orbit—the first time anyone will have flown more than 400 miles from the Earth since 1972—appears unlikely before 2023. Orion/SLS has survived primarily by sustaining jobs at the agency's field centers and old-line aerospace contractors. The recently announced Trump administration objective of accelerating the project and restoring lunar landings is too new to be evaluated, but the fate of the two Bush pronouncements alone shows that skepticism is warranted.

Conclusions

More than a quarter century has passed since the end of the Cold War and human spaceflight has continued with an ever-growing number of participants. Five new spacecraft are being developed in the United States—SpaceShipTwo, New Shepard, Crew Dragon, CST-100, and Orion—and another in Russia, the Federatsiya (Federation), to replace Soyuz. The Shenzhou will keep flying, as will the ISS (at least until the mid-2020s), and Europeans, Japanese, Canadians, and others will be on some or all of these vehicles. (ESA is funding the service module for Orion based on its ISS cargo ship.) The lack of any compelling, nonpolitical purpose for human spaceflight, and the failure to fulfill any of the astrofuturist expectations for Moon bases, Mars colonies, and the like, have not stopped the activity.

As we have seen, that is largely because it remains the ultimate signal of a major power's status and technoscientific capability, plus it sustains jobs in places and industries that grew explosively during the missile and space races. Moreover, the global astroculture continues to generate visions of human deep-space exploration and colonization, and the public response to astronaut experiences and space tourism offers ample evidence that many want to go into space, and not just see it through the eyes of robots. But the years since the end of the Cold War—indeed since the end of Apollo—have demonstrated how difficult and expensive human spaceflight remains, even without the serious health challenges posed by zero gravity and cosmic radiation during extended stays in space. Unless there is a new geopolitically inspired race, or a discovery of extraterrestrial life, it is hard to imagine that the situation will change quickly, even if we do go back to the Moon in the 2020s.

Global astroculture continues to generate visions of human deep-space exploration and colonization, and the public response to astronaut experiences and space tourism offers ample evidence that many want to go into space, and not just see it through the eyes of robots.

EPILOGUE: THE PAST AND FUTURE OF SPACEFLIGHT

Spaceflight arrived with astonishing rapidity in the mid-twentieth century. Only twenty-seven years after the first V-2 flights in 1942, humans landed on the Moon. By 1989, robotic spacecraft had flown by every major planet, and four were headed into interstellar space. Those achievements would not have been possible without the insights and advocacy of the earliest theorists and promoters, magnified by an astroculture that made spaceflight seem feasible and exciting to a much larger audience. But without the driving forces of war, international arms races, and political competition, spaceflight would have taken much longer to emerge and would have taken an entirely different course.

Once the Cold War space race slackened and then died, the pace of change, particularly in human spaceflight, decelerated. Yet the power of space achievements to signal

By 1989, robotic spacecraft had flown by every major planet, and four were headed into interstellar space.

the importance and technological capability of nations helped sustain the programs of the original space powers, and drew in new nations on every continent. Even as spaceflight globalized, it accelerated the globalization of the world, notably through the spread of television and entertainment via communications satellites. The sheer usefulness of space infrastructure, whether for making money, bolstering military power, enabling navigation, predicting weather, or providing warnings, meant that space activity would have expanded even without the advantages of geopolitical signaling. Moreover, as I have noted throughout the book, the institutionalization of spaceflight in government agencies, corporations, universities, and research centers created jobs and technoscientific capability that sustained political support for space programs, notably in sectors with less obvious use value (but often greater signaling value) like human spaceflight, planetary exploration, and space astronomy. Given all of these factors, it seems likely that spaceflight will continue to expand and globalize, particularly when it comes to Earth-orbiting infrastructure, which currently constitutes the great majority of everything we do in space.

However, the stability of our space infrastructure faces two major threats: space junk and space war. The number of dead satellites, rocket stages, and random detritus is already a problem, particularly in low Earth orbit. New constellations of hundreds and even thousands of

The stability of our space infrastructure faces two major threats: space junk and space war.

micro-spacecraft for communications and Earth observation are currently under construction. That could lead to the Kessler syndrome: a cascade of collisions creating clouds of debris that could make certain orbital zones unusable. Physical attacks on satellites, almost certainly as part of a war on Earth, could also trigger a cascade. They would certainly produce a new arms race in space.

As for the long term, the midcentury astrofuturists laid out an agenda that many still find persuasive: a human spaceflight–centered vision of Moon and Mars colonies and activity across the solar system. Space enthusiasts have been repeatedly disappointed that this reality has failed to materialize. There seems to be no reason to expect that it will any time soon either, although there will be human flights to the Moon and maybe Mars in the next two or three decades. Such enterprises are very expensive; sustaining political support over the long haul remains difficult if they cannot pay for themselves. There are also serious questions about the adaptability of human bodies to deep space radiation and low gravity. Space historians Roger Launius and Howard McCurdy have asked whether the intelligent machines or cyborgs we may create would be better suited to the task than fragile humans. Of course, we may not take well to being replaced.[1]

Another question is whether we can sustain deep space exploration if, as seems likely, humanity faces a severe global crisis in this century because of sea level rise, extreme weather, loss of arable land, and population growth, resulting in massive refugee flows and starvation.[2] Climate change is insidious because it works over timescales longer than our political systems seem designed to master. Robotic and human exploration may well seem like a disposable luxury in such a crisis, although advocates have argued that we need space colonies as an insurance policy against the destruction of our home planet. Space imagery has certainly helped foster a new planetary consciousness, one that may produce a sense of Earth's vulnerability. Satellites have also been scientifically central to understanding our changing global environment, information that is essential if we are to take effective action. And the science and technology developed for spaceflight have been and will be critical to shaping a response to climate change and energy conversion—for example, solar panels began as space technology. The problem is not finding technical solutions to mitigate the coming crises; it is finding the political will to do something effective.

Historians are not normally in the business of prediction. At best it is a risky proposition. But we can look back at the history of spaceflight and see human accomplishments that seemed impossible only a few decades earlier. It gives hope that we can solve our problems after all.

GLOSSARY

Black-powder rocket
The earliest and, for eight hundred years, only form of the rocket. Invented in China around 1100 CE, black powder was a slower-burning variant of gunpowder, which is a mixture of saltpeter, sulfur, and charcoal.

European Space Agency (ESA)
Founded in 1975 as a merger of the European Space Research Organization (ESRO) and the European Launcher Development Organization (ELDO), ESA is a cooperative organization that is independent of the European Union. It is dominated by Western European countries, but after the collapse of the Soviet bloc, it expanded to include Eastern European nations.

Geostationary Earth orbit (GEO)
A spacecraft orbiting at about 22,300 miles has a period of twenty-four hours, matching Earth's rotation. When the orbit is circular and the inclination is zero, that is, exactly over the equator, the satellite appears to hover over a location on the equator, and is thus geostationary. This orbit is primarily used by global communications and observation satellites.

Globalization
The integration of different world regions, societies, cultures, economies, and political systems.

Global Positioning System (GPS)
A system of satellites operated by the U.S. Air Force. These spacecraft are in twelve-hour orbits at about 11,000 miles altitude, have extremely precise atomic clocks, and transmit a time signal and information about their orbits. Receivers on Earth can calculate very accurate positions and altitudes by triangulating the signals from at least three satellites.

International Geophysical Year (IGY)
International scientific campaign focusing primarily on Earth's polar regions, atmosphere, oceans, magnetic field, ionosphere and near space, July 1, 1957–December 31, 1958.

Kessler syndrome
In 1978, NASA scientist Donald J. Kessler described a potential cascade of collisions among spacecraft and space junk, leading to orbital zones that could be unusable because of the density of the debris.

Liquid oxygen (LOX)
The most common oxidizer used in liquid-propellant rockets, oxygen liquefies at -297 degrees Fahrenheit at sea-level atmospheric pressure.

Liquid-propellant rocket
A rocket propulsion system based on the chemical reaction of one or more liquid propellants. Typically, there are two: a fuel and an oxidizer that will burn in a triggered or spontaneous combustion, producing a rapidly expanding jet of hot gas.

Low Earth orbit (LEO)
The LEO region is usually defined as 100–1,200 miles. Objects orbiting near Earth have period from ninety minutes to a few hours. The orbits of objects in the lowest part of the zone decay quickly due to friction with the tenuous, extreme outer atmosphere of Earth.

Medium Earth orbit (MEO)
An orbital region at about 11,000–12,000 miles altitude, occupied primarily by navigation spacecraft. These satellites in circular orbits at this altitude have periods of around twelve hours.

National Aeronautics and Space Administration (NASA)
U.S. civilian space agency founded in 1958 as an expansion of the National Advisory Committee for Aeronautics (NACA).

National Reconnaissance Office (NRO)
U.S. military space agency created in 1961 to carry out the development and production of reconnaissance satellites.

Planetization
The rising consciousness that Earth is a planet in the solar system like other planets. It is furthered by the availability of space images of Earth.

Rocket
A self-contained propulsion system in which all propellants are carried internally and force is created through the expulsion of a jet of hot gas or ionized plasma. It operates according to Newton's third law of motion: for every action, there is an equal and opposite reaction. The term is also used to describe a vehicle propelled by such a system. See also *black-powder rocket*, *liquid-propellant rocket*, and *solid-propellant rocket*.

Roscosmos
Russian government space agency founded in 1992. In 2015, reorganized as Roscosmos State Corporation for Space Activities.

Solid-propellant rocket
A rocket motor in which the propellants are mixed into a solid matrix. Its earliest form was the *black-powder rocket*. Modern solid propellants are complex mixtures containing rubber-like compounds mixed with oxygen-bearing perchlorates and powered aluminum. They burn from inside out along a lengthwise channel that is shaped to regulate the combustion rate.

NOTES

1 Spaceflight Dreams and Military Imperatives

1. Asif A. Siddiqi, *The Red Rockets' Glare: Spaceflight and the Soviet Imagination, 1857–1957* (Cambridge: Cambridge University Press, 2010), 18–30; James T. Andrews, *Red Cosmos: K. E. Tsiolkovskii, Grandfather of Soviet Rocketry* (College Station: Texas A&M University Press, 2009). For an unflattering view, see Michael Hagemeister, "The Conquest of Space and the Bliss of the Atoms: Konstantin Tsiolkovskii," in *Soviet Space Culture: Cosmic Enthusiasm in Socialist Societies*, edited by by Eva Maurer, Julia Richers, Monica Rüthers, and Carmen Scheide (Basingstoke, UK: Palgrave Macmillan, 2011), 27–41.

2. Tom D. Crouch, *Aiming for the Stars: The Dreamers and Doers of the Space Age* (Washington, DC: Smithsonian Institution Press, 1999); Christopher Gainor, *To a Distant Day: The Rocket Pioneers* (Lincoln: University of Nebraska Press, 2008).

3. David A. Clary, *Rocket Man: Robert H. Goddard and the Birth of the Space Age* (New York: Hyperion, 2003). The only Oberth biography available in English is a translation from the German of a book originally published in Russian: Boris V. Rauschenbach, *Hermann Oberth: The Father of Space Flight* (Clarence, NY: West-Art Press, 1994). It is not critical or scholarly.

4. *A Method* was republished in Robert H. Goddard, *Rockets* (New York: American Rocket Society, 1946). On its impact, see Frank H. Winter, "The Silent Revolution: How R. H. Goddard Helped Start the Space Age," in *History of Rocketry and Astronautics: Proceedings of the Thirty-Eighth History Symposium of the International Academy of Astronautics, Vancouver, British Columbia, Canada, 2004*, edited by Å. Ingemar Skoog (San Diego: Univelt, Inc., 2011), 3–54.

5. Michael J. Neufeld, "Weimar Culture and Futuristic Technology: The Rocketry and Spaceflight Fad in Germany, 1923–1933," *Technology and Culture* 31 (October 1990), 725–752.

6. Asif A. Siddiqi, "Deep Impact: Robert Goddard and the Soviet 'Space Fad' of the 1920s," *History and Technology* 20 (June 2004): 97–113.

7. Neufeld, "Weimar Culture"; Jared S. Buss, *Willy Ley: Prophet of the Space Age* (Gainesville: University Press of Florida, 2017), 25–55. Still valuable as an overview is Frank H. Winter, *Prelude to the Space Age: The Rocket Societies: 1924–1940* (Washington, DC: Smithsonian Institution Press, 1983).

8. Michael J. Neufeld, *Von Braun: Dreamer of Space, Engineer of War* (New York: Alfred A. Knopf, 2007), 7–48.

9. Tom D. Crouch, *Rocketeers and Gentlemen Engineers: A History of the American Institute of Aeronautics and Astronautics ... and What Came Before* (Reston, VA: AIAA, 2006), 25–52.
10. Clary, *Rocket Man*; J. D. Hunley, "The Enigma of Robert H. Goddard," *Technology and Culture* 36 (April 1995): 327–350; Alexander MacDonald, *The Long Space Age: The Economic Origins of Space Exploration from Colonial America to the Cold War* (New Haven: Yale University Press, 2017), 105–159.
11. Michael J. Neufeld, *The Rocket and the Reich: Peenemünde and the Coming of the Ballistic Missile Era* (New York: The Free Press, 1995).
12. Michael J. Neufeld, "Hitler, the V-2, and the Battle for Priority, 1939–1943," *Journal of Military History* 57 (July 1993): 511–538.
13. Michael J. Neufeld, "Wernher von Braun, the SS and Concentration Camp Labor: Questions of Moral, Political and Criminal Responsibility," *German Studies Review* 25 (February 2002): 57–78.
14. Neufeld, *The Rocket and the Reich*; Jens-Christian Wagner, *Produktion des Todes: Das KZ Mittelbau-Dora* (Göttingen: Wallstein, 2001); André Sellier, *A History of the Dora Camp* (Chicago: Ivan Dee, 2003).
15. Frank H. Winter, *America's First Rocket Company: Reaction Motors, Inc.* (Reston, VA: AIAA, 2017); Clayton R. Koppes, *JPL and the American Space Program: A History of the Jet Propulsion Laboratory* (New Haven: Yale University Press, 1982).
16. Siddiqi, *The Red Rockets' Glare*, 155–195.
17. Neufeld, *The Rocket and the Reich*, 267–279.
18. Asif A. Siddiqi, *Challenge to Apollo: The Soviet Union and the Space Race, 1945–1974* (Washington, DC: NASA, 2000); Boris Chertok, *Rockets and People*, vol. 1 (Washington, DC: NASA, 2005).
19. Neufeld, *Von Braun*, 199–222; Brian Crim, *Our Germans: Project Paperclip and the National Security State* (Baltimore: Johns Hopkins University Press, 2018).
20. Michael J. Neufeld, "The Nazi Aerospace Exodus: Towards a Global, Transnational History," *History and Technology* 28 (2012): 49–67; Olivier Huwart, *Du V2 à Veronique: La naissance des fusées françaises* (Rennes: Marines éditions, 2004).
21. Siddiqi, *The Red Rockets' Glare*, and his *Challenge to Apollo*.
22. David H. DeVorkin, *Science with a Vengeance: How the Military Created the US Space Sciences after World War II* (New York: Springer-Verlag, 1992); J. D. Hunley, *The Development of Propulsion Technology for U.S. Space-Launch Vehicles, 1926–1991* (College Station: Texas A&M University Press, 2007). For a popular history that gives the air force and its contractors due credit, see T.

A. Heppenheimer, *Countdown: A History of Space Flight* (New York: John Wiley & Sons, 1997).

23. Christopher Gainor, *The Bomb and America's Missile Age* (Baltimore: Johns Hopkins University Press, 2018); Jacob Neufeld, *The Development of Ballistic Missiles in the United States Air Force, 1945–1960* (Washington, DC: Office of Air Force History, 1990).

24. Howard E. McCurdy, *Space and the American Imagination* (Washington, DC: Smithsonian Institution Press, 1997), 29–51; Siddiqi, *The Red Rockets' Glare*, 290–331.

2 The Cold War Space Race

1. Michael J. Neufeld, "Orbiter, Overflight and the First U.S. Satellite: New Light on the Vanguard Decision," in *Reconsidering Sputnik*, edited by Roger D. Launius, John M. Logsdon, and Robert W. Smith (Amsterdam: Harwood Academic Publishers, 2000), 231–257.

2. Allan A. Needell, *Science, Cold War and the American State: Lloyd V. Berkner and the Balance of Professional Ideals* (Amsterdam: Harwood Academic, 2000), 297–353.

3. Walter A. McDougall, *... the Heavens and the Earth: A Political History of the Space Age* (New York: Basic Books, 1985), 112–134; R. Cargill Hall, "The Eisenhower Administration and the Cold War: Framing American Astronautics to Serve National Security," *Prologue* 27 (Spring 1995): 58–72.

4. Siddiqi, *The Red Rockets' Glare*, 313–324.

5. Neufeld, "Orbiter, Overflight."

6. Siddiqi, *The Red Rockets' Glare*, 324–335.

7. Kim McQuaid, "Sputnik Reconsidered: Image and Reality in the Early Space Age," *Canadian Review of American Studies* 37 (2007): 371–401; McDougall, *... the Heavens*, 141–156.

8. Siddiqi, *Challenge to Apollo*, 167–174.

9. McDougall, *... the Heavens*, 141–156; Neufeld, *Von Braun*, 311–323.

10. Michael J. Neufeld, "The End of the Army Space Program: Interservice Rivalry and the Transfer of the Von Braun Group to NASA, 1958–1959," *Journal of Military History* 69 (July 2005): 737–758.

11. On applying economic signaling theory to the space race, see MacDonald, *The Long Space Age*, 7–11, 160–206.

12. Dwayne A. Day, John M. Logsdon, and Brian Latell, eds., *Eye in the Sky: The Story of the Corona Spy Satellites* (Washington, DC: Smithsonian Institution Press, 1998); James E. David, *Spies and Shuttles: NASA's Secret Relationships with the DoD and CIA* (Gainesville: University Press of Florida, 2015).

13. Margaret A. Weitekamp, *Right Stuff, Wrong Sex: America's First Women in Space Program* (Baltimore, MD: Johns Hopkins University Press, 2004).
14. John M. Logsdon, *John F. Kennedy and the Race to the Moon* (New York: Palgrave Macmillan, 2010); Michael R. Beschloss, "Kenney and the Decision to Go to the Moon," in *Spaceflight and the Myth of Presidential Leadership*, edited by Roger D. Launius and Howard E. McCurdy (Urbana: University of Illinois Press, 1997), 51–67.
15. A readable history of Mercury, Gemini, and Apollo from the engineers' point of view is Charles Murray and Catherine Bly Cox, *Apollo: The Race to the Moon* (New York: Simon & Schuster, 1989).
16. On the Soviet program in the sixties, see Siddiqi, *Challenge to Apollo*.
17. John M. Logsdon, *After Apollo? Richard Nixon and the American Space Program* (New York: Palgrave Macmillan, 2015); Joan Hoff, "The Presidency, Congress, and the Deceleration of the U.S. Space Program in the 1970s," in Launius and McCurdy, *Spaceflight and the Myth*, 92–132.
18. Asif A. Siddiqi, "Soviet Space Power during the Cold War," in *Harnessing the Heavens: National Defense through Space*, edited by Paul G. Gillespie and Grant T. Weller (Chicago: Imprint Publications, 2008), 135–150.
19. Angelina Callahan, "The Origins and Flagship Project of NASA's International Program: The Ariel Case Study," in *NASA Spaceflight: A History of Innovation*, edited by Roger D. Launius and Howard E. McCurdy (Chur: Palgrave Macmillan, 2017), 33–55; Andrew B. Godefroy, *Defence and Discovery: Canada's Military Space Program, 1945–74* (Vancouver: UBC Press, 2011); J. Krige, A. Russo, and L. Sebesta, *A History of the European Space Agency 1958–1987*, 2 vols. (Noordwijk: ESA, 2000); Iris Chang, *Thread of the Silkworm* (New York: Basic Books, 1995); Gregory Kulacki and Jeffrey G. Lewis, *A Place for One's Mat: China's Space Program, 1956–2003* (Cambridge, MA: American Academy of Arts and Sciences, 2009), https://www.amacad.org/publications/spaceChina.pdf, accessed November 22, 2017.
20. Roger D. Launius, *Space Stations: Base Camps to the Stars* (Washington, DC: Smithsonian Books, 2003).
21. Michael J. Neufeld, "The 'von Braun Paradigm' and NASA's Long-Term Planning for Human Spaceflight," in *NASA's First 50 Years: Historical Perspectives*, edited by Steven J. Dick (Washington, DC: NASA, 2010), 325–347; Lyn Ragsdale, "Politics Not Science: The U.S. Space Program in the Reagan and Bush Years," in Launius and McCurdy, *Spaceflight and the Myth*, 133–171, esp. 156–161.
22. Logsdon, *After Apollo?*, 143–301; Heppenheimer, *Countdown*, 305–328.

23. John M. Logsdon, "Selling the Space Shuttle: Early Developments," in Launius and McCurdy, *NASA Spaceflight*, 185–214. For a study of how NASA and the media framed the shuttle and space station, see Valerie Neal, *Spaceflight in the Shuttle Era and Beyond: Redefining Humanity's Purpose in Space* (New Haven: Yale University Press, 2017).
24. Frances FitzGerald, *Way Out There in the Blue: Reagan, Star Wars and the End of the Cold War* (New York: Simon & Schuster, 2000).

3 Space Science and Exploration

1. Paul Ceruzzi, "An Unforeseen Revolution: Computers and Expectations, 1935–1985," in *Imagining Tomorrow: History, Technology, and the American Future*, edited by Joseph J. Corn (Cambridge, MA: MIT Press, 1986), 188–201; Roger D. Launius and Howard E. McCurdy, *Robots in Space: Technology, Evolution, and Interplanetary Travel* (Baltimore: Johns Hopkins University Press, 2008).
2. DeVorkin, *Science with a Vengeance*.
3. Abigail Foerstner, *James Van Allen: The First Eight Billion Miles* (Iowa City: University of Iowa Press, 2007).
4. Wesley T. Huntress Jr. and Mikhail Ya. Marov, *Soviet Robots in the Solar System: Mission Technologies and the Discoveries* (Chichester, UK: Springer Praxis, 2011), 67–142; Edward Clinton Ezell and Linda Neuman Ezell, *On Mars: Exploration of the Red Planet, 1958–1978* (Washington, DC: NASA, 1984), 25–50.
5. Koppes, *JPL*, 113–133, 161–184.
6. Needell, *Science*, 155–162.
7. William David Compton, *Where No Man Has Gone Before: A History of Apollo Lunar Exploration Missions* (Washington, DC: NASA, 1989).
8. Huntress and Marov, *Soviet Robots*, 21–25.
9. Ibid., 143–366.
10. Robert S. Kraemer, *Beyond the Moon: A Golden Age of Planetary Exploration, 1971–1978* (Washington, DC: Smithsonian Institution Press, 2000); Ezell and Ezell, *On Mars*; W. Henry Lambright, *Why Mars: NASA and the Politics of Space Exploration* (Baltimore: Johns Hopkins University Press, 2014), 17–69.
11. DeVorkin, *Science with a Vengeance*.
12. David DeVorkin, "The Space Age and Disciplinary Change in Astronomy," in Dick, *NASA's First 50 Years*, 389–426; Robert W. Smith, "The Making of Space Astronomy: A Gift of the Cold War," in *Earth-Bound to Satellite: Telescopes, Skills and Networks*, edited by A. D. Morrison-Low, Sven Dupre, Stephen Johnston, and Giorgio Strano (Leiden: Brill, 2011), 235–249.

13. David H. DeVorkin, *Fred Whipple's Empire: The Smithsonian Astrophysical Observatory, 1955–1973* (Washington, DC: Smithsonian Institution Scholarly Press, 2018).
14. Robert W. Smith, *The Space Telescope: A Study of NASA, Science, Technology and Politics*, revised edition with a new afterword (Cambridge: Cambridge University Press, 1993); W. Henry Lambright, "Big Science in Space: Viking, Cassini, and the Hubble," in *Exploring the Solar System: The History and Science of Planetary Exploration*, edited by Roger D. Launius (New York: Palgrave Macmillan, 2013), 129–148.
15. Karl Hufbauer, *Exploring the Sun: Solar Science since Galileo* (Baltimore: Johns Hopkins University Press, 1991), 160–312.
16. Smith, "The Making of Space Astronomy."
17. Steven J. Dick, *Life on Other Worlds: The 20th-Century Extraterrestrial Life Debate* (Cambridge: Cambridge University Press, 1998), 169–199.
18. See the contributions of Edward S. Goldstein, James R. Fleming, and Erik M. Conway in Dick, *NASA's First 50 Years*, 503–585, and those of Erik M. Conway, Andrew K. Johnston, and Roger D. Launius in Launius, *Exploring the Solar System*, 183–243.
19. John M. Logsdon, "The Survival Crisis of the US Solar System Exploration Program in the 1980s," in Launius, *Exploring the Solar System*, 45–76.
20. On the mirror flaw, see the afterword in Smith, *The Space Telescope*; Roger D. Launius and David H. DeVorkin, eds., *Hubble's Legacy: Reflections by Those Who Dreamed It, Built It, and Observed the Universe with It* (Washington, DC: Smithsonian Institution Scholarly Press, 2014), http://opensi.si.edu/index.php/smithsonian/catalog/book/57, accessed November 22, 2017.
21. Huntress and Marov, *Soviet Robots*, 367–405.
22. Howard E. McCurdy, *Faster, Better, Cheaper: Low-Cost Innovation in the U.S. Space Program* (Baltimore: Johns Hopkins University Press, 2001); Peter J. Westwick, *Into the Black: JPL and the American Space Program, 1976–2004* (New Haven: Yale University Press, 2007).
23. Michael J. Neufeld, "Transforming Solar System Exploration: The Origins of the Discovery Program, 1989–1993," *Space Policy* 30 (2014): 5–12; Erik M. Conway, *Exploration and Engineering: The Jet Propulsion Laboratory and the Quest for Mars* (Baltimore: Johns Hopkins University Press, 2015), 87–139.
24. Conway, *Exploration*, 140–343; Michael J. Neufeld, "The Discovery Program: Competition, Innovation, and Risk in Planetary Exploration," in Launius and McCurdy, *NASA Spaceflight*, 267–290, and my "First Mission to Pluto: Policy, Politics, Science and Technology in the Origins of New Horizons, 1989–2003," *Historical Studies in the Natural Sciences* 44 (2014): 234–276.

25. Arturo Russo, "Parachuting onto Another World: The European Space Agency's Huygens Mission to Titan," in Launius, *Exploring the Solar System*, 275–321, and his "Europe's Path to Mars: The European Space Agency's Mars Express Mission," *Historical Studies in the Natural Sciences* 41 (2011): 123–178; Patrick Besha, "Policy Making in China's Space Program: A History and Analysis of the Chang'e Lunar Orbiter Project," *Space Policy* 26 (2010): 214–221.

4 A Global Space Infrastructure

1. Jürgen Osterhammel and Neils P. Petersson, *Globalization: A Short History* (Princeton: Princeton University Press, 2005).

2. Day, Logsdon, and Latell, *Eye in the Sky*; Jeffrey T. Richelson, *America's Secret Eyes in Space: The U.S. Keyhole Spy Satellite Program* (New York: Harper & Row. 1990); contributions by Henry R. Hertzfeld and Ray A. Williamson; Erik M. Conway; David J. Whalen; and W. Henry Lambright in *Societal Impact of Spaceflight*, edited by Steven J. Dick and Roger D. Launius (Washington, DC: NASA, 2007), 237–330.

3. Peter Gorin, "ZENIT: The Soviet Response to CORONA," in Day, Logsdon, and Latell, *Eye in the Sky*, 157–170; Siddiqi, "Soviet Space Power"; Bart Hendrickx, "A History of Soviet/Russian Meteorological Satellites," *JBIS Space Chronicle* 57 (2004), suppl. 1: 56–102.

4. Erik M. Conway, "Satellites and Security: Space in Service to Humanity," in Dick and Launius, *Societal Impact*, 267–288, and his *Atmospheric Science at NASA: A History* (Baltimore: Johns Hopkins University Press, 2008).

5. Jeffrey T. Richelson, *America's Space Sentinels: The History of the DSP and SBIRS Satellite Systems*, 2nd ed. (Lawrence: University Press of Kansas, 2012); Pavel Podvig, "History and the Current Status of the Russian Early-Warning System," *Science and Global Security*10 (2002): 21–60, http://russianforces.org/podvig/2002/03/history_and_the_current_status.shtml, accessed November 22, 2017.

6. Declassified GAMBIT and HEXAGON official histories and information are available at http://www.nro.gov/history/csnr/gambhex/,accessed November 22, 2017.

7. Richelson, *America's Secret Eyes*.

8. Asif A. Siddiqi, "Staring at the Sea: The Soviet Rorsat and Eorsat Programmes," *JBIS* 52 (1999): 397–416.

9. On the impact of satellite-enabled global transparency on the Cold War, see John Lewis Gaddis, "The Long Peace: Elements of Stability in the Postwar International System," *International Security* 10 (4) (Spring 1986): 99–142.

10. Roger D. Launius, "Global Instantaneous Telecommunications and the Development of Satellite Technology," in Launius and McCurdy, *NASA Spaceflight*, 57–87.
11. On advocacy for deploying weapons, and on comsats in the U.S. military, see the contributions of Everett C. Dolman, "Astropolitics and *Astropolitik* Strategy and Space Deployment," and Rick W. Sturdevant, "Giving Voice to Global Reach, Global Power: Satellite Communications in U.S. Military Affairs, 1966–2007," respectively, in Gillespie and Weller, *Harnessing the Heavens*, 111–133 and 191–213.
12. Martin J. Collins, *A Telephone for the World: Iridium, Motorola, and the Making of a Global Age* (Baltimore: Johns Hopkins University Press, 2018).
13. Paul Ceruzzi, *GPS* (Cambridge, MA: MIT Press, 2018); Richard D. Easton and Eric F. Frazier, *GPS Declassified: From Smart Bombs to Smartphones* (n.p.: Potomac Books, 2013).
14. Ceruzzi, *GPS*; Rick W. Sturdevant, "NAVSTAR, the Global Positioning System: A Sampling of Its Military, Civil, and Commercial Impact," in Dick and Launius, *Societal Impact*, 331–351.
15. Satellite Industry Association, "State of the Satellite Industry Report," May 2012, https://www.sia.org/wp-content/uploads/2012/05/FINAL-2012-State-of-Satellite-Industry-Report-20120522.pdf, accessed November 24, 2017.
16. Dean Cheng, "The Long March Upward: A Review of China's Space Program," in Gillespie and Weller, *Harnessing the Heavens*, 151–163.

5 Astroculture

1. Alexander C. T. Geppert, "European Astrofuturism, Cosmic Provincialism; Historicizing the Space Age," in *Imagining Outer Space: European Astroculture in the Twentieth Century*, edited by Alexander C. T. Geppert (Basingstoke, UK: Palgrave Macmillan, 2012), 8.
2. Brian W. Aldiss, with David Wingrove, *Trillion Year Spree: The History of Science Fiction* (New York: Atheneum, 1986); David Seed, *Science Fiction: A Very Short Introduction* (Oxford: Oxford University Press, 2013).
3. Winter, "The Silent Revolution"; Koppes, *JPL*, 8, 19.
4. DeWitt Douglas Kilgore, *Astrofuturism: Science, Race, and Visions of Utopia in Space* (Philadelphia: University of Pennsylvania Press, 2003), 2; McCurdy, *Space and the American Imagination*, 29–51.
5. Siddiqi, *The Red Rockets' Glare*, 290–313.
6. Steven J. Dick, *Plurality of Worlds: The Origins of the Extraterrestrial Life Debate from Democritus to Kant* (Cambridge: Cambridge University Press, 1982);

Michael J. Crowe, *The Extraterrestrial Life Debate, 1750–1900: The Idea of the Plurality of Worlds from Kant to Lowell* (Cambridge: Cambridge University Press, 1986).

7. Robert Markley, *Dying Planet: Mars in Science and the Imagination* (Durham, NC: Duke University Press, 2005); K. Maria D. Lane, *Geographies of Mars: Seeing and Knowing the Red Planet* (Chicago: University of Chicago Press, 2011).

8. Dick, *Life on Other Worlds*, 137–168; Greg Eghigian, "'A Transatlantic Buzz': Flying Saucers, Extraterrestrials, and America in Postwar Germany," *Journal of Transatlantic Studies*, 12 (2014): 282–303; Alexander C. T. Geppert, "Extraterrestrial Encounters: UFOs, Science and the Quest For Transcendence, 1947–1972," *History and Technology* 28 (September 2012): 335–362.

9. McCurdy, *Space and the American Imagination*, 109–137; Dick, *Life on Other Worlds*, 53–65.

10. Dick, *Life on Other Worlds*, 200–235.

11. Michael J. Neufeld, ed., *Spacefarers: Images of Astronauts and Cosmonauts in the Heroic Age of Spaceflight* (Washington, DC: Smithsonian Institution Scholarly Press, 2013), especially the contributions of Margaret A. Weitekamp, Matthew H. Hersch, James Spiller, Andrew Jenks, and Trevor S. Rockwell.

12. Matthew H. Hersch, "'Capsules Are Swallowed': The Mythology of the Pilot in American Spaceflight," in Neufeld, *Spacefarers*,, 35–55, and his *Inventing the American Astronaut* (New York: Palgrave Macmillan, 2012).

13. James T. Andrews and Asif A. Siddiqi, eds., *Into the Cosmos: Space Exploration and Soviet Culture* (Pittsburgh: University of Pittsburgh Press, 2011), especially Asif A. Siddiqi, "Cosmic Contradictions, Popular Enthusiasm and Secrecy in the Soviet Space Program," 47–76, and Slava Gerovitch, "The Human inside a Propaganda Machine: The Public Image and Professional Identity of Soviet Cosmonauts," 77–106.

14. Slava Gerovitch, *Soviet Space Mythologies: Public Images, Private Memories, and the Making of a Cultural Identity* (Pittsburgh: University of Pittsburgh Press, 2015); Andrew Jenks, "The Sincere Deceiver: Yuri Gagarin and the Search for a Higher Truth," in Andrews and Siddiqi, *Into the Cosmos*, 107–132; Andrew L. Jenks, *The Cosmonaut Who Wouldn't Stop Smiling: The Life and Legend of Yuri Gagarin* (Dekalb: Northern Illinois University Press, 2012).

15. Maurer, Richers, Rüthers and Scheide, *Soviet Space Culture*, especially Asif A. Siddiqi, "From Cosmic Enthusiasm to Nostalgia for the Future: A Tale of Soviet Space Culture," 283–306; Andrews and Siddiqi, *Into the Cosmos*.

16. Matthew D. Tribbe, *No Requiem for the Space Age: The Apollo Moon Landings and American Culture* (New York: Oxford University Press, 2014); Neal, *Spaceflight in the Shuttle Era*, 63–98. For the best biography of an astronaut,

see James R. Hansen, *First Man: The Life of Neil A. Armstrong* (New York: Simon & Schuster, 2005).

17. Weitekamp, *Right Stuff, Wrong Sex*; Neil M. Maher, *Apollo in the Age of Aquarius* (Cambridge, MA: Harvard University Press, 2017), 10–53, 137–182.

18. Jennifer Ross-Nazzal, "You've Come a Long Way, Maybe: The First Six Women Astronauts and the Media," in Neufeld, *Spacefarers*, 175–201; Amy E. Foster, *Integrating Women into the Astronaut Corps: Politics and Logistics at NASA, 1972–2004* (Baltimore: The Johns Hopkins University Press, 2011).

19. Roshanna P. Silvester, "She Orbits over the Sex Barrier: Soviet Girls and the Tereshkova Moment," in Andrews and Siddiqi, *Into the Cosmos*, 195–212; Neal, *Spaceflight in the Shuttle Era*, 83–98.

20. Benjamin Lazier, "Earthrise, or the Globalization of the World Picture," *American Historical Review* 116 (June 2011), 602–30; Robert Poole, *Earthrise: How Man First Saw* the Earth (New Haven: Yale University Press, 2008); Maher, *Apollo in the Age of Aquarius*, 93–136.

21. Alexander C. T. Geppert, "Where the Beyond Begins: Pierre Teilhard de Chardin and the Spatialization of Space ... ," unpublished article, courtesy Alexander Geppert.

22. Steven J. Dick, "Space, Time and Aliens: The Role of the Imagination in Outer Space," in Geppert, *Imagining Outer Space*, 27–44; Steven J. Dick and Mark L. Lupisella, *Cosmos and Culture: Cultural Evolution in a Cosmic Context* (Washington, DC: NASA, 2009); Keay Davidson, *Carl Sagan: A Life* (New York: John Wiley & Sons, 1999).

23. Elizabeth A. Kessler, *Picturing the Cosmos: Hubble Space Telescope Images and the Astronomical Sublime* (Minneapolis: University of Minnesota Press, 2012).

24. Teasel Muir-Harmony, "Selling Space Capsules, Moon Rocks, and America: Spaceflight in U.S. Public Diplomacy, 1961–1979," in *Reasserting America in the 1970s*, edited by Hallvard Notiker, Giles Scott-Smith, and David J. Snyder (Manchester, UK: Manchester University Press, 2016), 127–142; Maurer, Richers, Rüthers, and Scheide, *Soviet Space Culture*, 167–225.

25. James R. Hansen, "The *Taikonaut* as Icon: The Cultural and Political Significance of Yang Liwei, China's First Space Traveler," in Dick and Launius, *Societal Impact*, 103–117.

6 Human Spaceflight after the Cold War

1. George H. W. Bush speech, July 20, 1989, http://www.presidency.ucsb.edu/ws/?pid=17321, accessed November 7, 2017; George W. Bush speech, January

14, 2004, https://www.nasa.gov/missions/solarsystem/bush_vision.html, accessed October 6, 2017; Neal, *Spaceflight in the Shuttle Era*, 176–190.

2. Michael Cassutt, "Secret Space Shuttles," *Air & Space Smithsonian*, August 2009, https://www.airspacemag.com/space/secret-space-shuttles-35318554/, accessed October 9, 2017.

3. Ragsdale, "Politics not Science," in Launius and McCurdy, *Spaceflight and the Myth*, 156–61; Neal, *Spaceflight in the Shuttle Era*, 134–162.

4. Thor Hogan, *Mars Wars: The Rise and Fall of the Space Exploration Initiative* (Washington, DC: NASA, 2007); Neufeld, "The 'von Braun Paradigm.'"

5. Marcia S. Smith, "NASA's Space Station Program: Evolution and Current Status, Testimony before the House Science Committee," April 4, 2001, https://history.nasa.gov/isstestimony2001.pdf, accessed October 15, 2017; Launius, *Space Stations*, 151–163.

6. Launius, *Space Stations*, 163–173.

7. Ibid., 175–194.

8. Diane Vaughan, *The Challenger Launch Decision: Risky Technology, Culture and Deviance at NASA* (Chicago: University of Chicago Press, 1996); Columbia Accident Investigation Board, *Report Volume 1* (Washington, DC: NASA, 2003).

9. Kulacki and Lewis, *A Place for One's Mat*, 19–29; Hansen, "The *Taikonaut* as Icon."

10. Chris Dubbs and Emiline Paat-Dahlstrom, *Realizing Tomorrow: The Path to Private Spaceflight* (Lincoln: University of Nebraska Press, 2011); Julian Guthrie, *How to Make a Spaceship: A Band of Renegades, an Epic Race and the Birth of Private Space Flight* (New York: Penguin Press, 2016)

11. Christian Davenport, *The Space Barons: Elon Musk, Jeff Bezos, and the Quest to Colonize the Cosmos* (New York: Public Affairs, 2018).

12. John M. Logsdon, "Encouraging New Space Firms," and W. Henry Lambright, "NASA, Industry, and the Commercial Crew Development Program: The Politics of Partnership," respectively, in Launius and McCurdy, *NASA Spaceflight*, 237–265 and 349–377.

13. Lambright, "NASA, Industry," 365–375.

14. Glen R. Asner and Stephen J. Garber, *Origins of 21st Century Spaceflight: A History of NASA's Decadal Planning Team and the Vision for Space Exploration, 1999–2004* (Washington, DC: NASA, 2018).

Epilogue

1. Launius and McCurdy, *Robots in Space*.

2. For a worst-case scenario in the form of a novella, see Naomi Oreskes and Erik M. Conway, *The Collapse of Western Civilization: A View from the Future* (New York: Columbia University Press, 2014).

FURTHER READINGS

Buss, Jared S. *Willy Ley: Prophet of the Space Age*. Gainesville, FL: University Press of Florida, 2017.

Ceruzzi, Paul. *GPS: A Concise History*. Cambridge, MA: MIT Press, 2018.

Clary, David A. *Rocket Man: Robert H. Goddard and the Birth of the Space Age*. New York: Hyperion, 2003.

Crouch, Tom D. *Aiming for the Stars: The Dreamers and Doers of Space Exploration*. Washington, DC: Smithsonian Institution Press, 1999.

DeVorkin, David H. *Science with a Vengeance: How the Military Created the US Space Sciences after World War II*. New York: Springer, 1992.

Dick, Steven J. *Life on Other Worlds: The 20th-Century Extraterrestrial Life Debate*. Cambridge: Cambridge University Press, 1998.

FitzGerald, Frances. *Way Out There in the Blue: Reagan, Star Wars and the End of the Cold War*. New York: Simon & Schuster, 2000.

Geppert, Alexander C. T., ed. *Imagining Outer Space: European Astroculture in the Twentieth Century*. Basingstoke, UK: Palgrave Macmillan, 2012.

Gerovitch, Slava. *Soviet Space Mythologies: Public Images, Private Memories, and the Making of a Cultural Identity*. Pittsburgh: University of Pittsburgh Press, 2015.

Hansen, James R. *First Man: The Life of Neil A. Armstrong*. New York: Simon & Schuster, 2005.

Heppenheimer, T. A. *Countdown: A History of Spaceflight*. New York: John Wiley & Sons, 1997.

Hersch, Matthew H. *Inventing the American Astronaut*. New York: Palgrave Macmillan, 2012.

Jenks, Andrew L. *The Cosmonaut Who Wouldn't Stop Smiling: The Life and Legend of Yuri Gagarin*. Dekalb, IL: Northern Illinois University Press, 2012.

Kilgore, DeWitt Douglas. *Astrofuturism: Science, Race, and Visions of Utopia in Space*. Philadelphia: University of Pennsylvania Press, 2003.

Kulacki, Gregory, and Jeffrey G. Lewis. *A Place for One's Mat: China's Space Program, 1956–2003*. Cambridge, MA: American Academy of Arts and Sciences, 2009. https://www.amacad.org/publications/spaceChina.pdf.

Launius, Roger D. *Space Stations: Base Camps to the Stars*. Washington, DC: Smithsonian Books, 2003.

Launius, Roger D., and Howard E. McCurdy. *Robots in Space: Technology, Evolution, and Interplanetary Travel*. Baltimore: Johns Hopkins University Press, 2008.

Logsdon, John M. *John F. Kennedy and the Race to the Moon*. New York: Palgrave Macmillan, 2010.

McCurdy, Howard E. *Space and the American Imagination*. Washington, DC: Smithsonian Institution Press, 1997.

McDougall, Walter A. *... the Heavens and the Earth: A Political History of the Space Age*. New York: Basic Books, 1985.

Murray, Charles, and Catherine Bly Cox. *Apollo: The Race to the Moon*. New York: Simon & Schuster, 1989.

Neal, Valerie. *Spaceflight in the Shuttle Era and Beyond: Redefining Humanity's Purpose in Space*. New Haven, CT: Yale University Press, 2017.

Neufeld, Michael J. *The Rocket and the Reich: Peenemünde and the Coming of the Ballistic Missile Era*. New York: The Free Press, 1995.

Neufeld, Michael J. *Von Braun: Dreamer of Space, Engineer of War*. New York: Alfred A. Knopf, 2007.

Siddiqi, Asif A. *Challenge to Apollo: The Soviet Union and the Space Race, 1945–1974*. Washington, DC: NASA, 2000. Republished as *Sputnik and the Soviet Space Challenge and The Soviet Space Race with Apollo*. Gainesville: University Press of Florida, 2003.

Siddiqi, Asif A. *The Red Rockets' Glare: Spaceflight and the Soviet Imagination, 1857–1957*. Cambridge: Cambridge University Press, 2010.

Smith, Robert W. *The Space Telescope: A Study of NASA, Science, Technology, and Politics*. Cambridge: Cambridge University Press, 1989.

Weitekamp, Margaret. *Right Stuff, Wrong Sex: America's First Women in Space Program*. Baltimore, MD: Johns Hopkins University Press, 2004.

Westwick, Peter J. *Into the Black: JPL and the American Space Program, 1976–2004*. New Haven, CT: Yale University Press, 2007.

INDEX

Note: Page numbers in italics indicate an illustration; page numbers in bold indicate a glossary definition.

DR. MICHAEL J. NEUFELD is a Senior Curator in the Space History Department at the Smithsonian's National Air and Space Museum. He is the author or editor of eight other books, most notably *Von Braun: Dreamer of Space, Engineer of War*.